U.S. Department
of Transportation
**National Highway
Traffic Safety
Administration**

DOT HS 810 580

March 2006

Analysis of Red Light Violation Data Collected from Intersections Equipped with Red Light Photo Enforcement Cameras

Research and
Innovative Technology
Administration

Volpe National
Transportation Systems Center,
Cambridge, MA 02142-1093

This document is available to the public through the National Technical Information Service, Springfield, VA 22161

> **NOTICE**
> This document is disseminated under the sponsorship of the Department of Transportation in the interest of information exchange. The United States Government assumes no liability for its contents or use thereof.

# REPORT DOCUMENTATION PAGE		*Form Approved* *OMB No. 0704-0188*

Public reporting burden for this collection of information is estimated to average one hour per response, including the time for reviewing instructions, searching existing data sources, gathering and maintaining the data needed, and completing and reviewing the collection of information. Send comments regarding this burden estimate or any other aspect of this collection of information, including suggestions for reducing this burden, to Washington Headquarters Services, Directorate for Information Operations and Reports, 1215 Jefferson Davis Highway, Suite 1204, Arlington, VA 22202-4302, and to the Office of Management and Budget, Paperwork Reduction Project (0704-0188), Washington, DC 20503.

1. AGENCY USE ONLY (Leave blank)	2. REPORT DATE March 2006	3. REPORT TYPE AND DATES COVERED Project Memorandum October 2003 – October 2005	
4. TITLE AND SUBTITLE Analysis of Red Light Violation Data Collected from Intersections Equipped with Red Light Photo Enforcement Cameras			5. FUNDING NUMBERS PPA # HS-19
6. AUTHOR(S) C. Y. David Yang and Wassim G. Najm			
7. PERFORMING ORGANIZATION NAME(S) AND ADDRESS(ES) U.S. Department of Transportation Research and Innovative Technology Administration John A. Volpe National Transportation Systems Center Cambridge, MA 02142			8. PERFORMING ORGANIZATION REPORT NUMBER DOT-VNTSC-NHTSA-05-01
9. SPONSORING/MONITORING AGENCY NAME(S) AND ADDRESS(ES) U.S. Department of Transportation National Highway Traffic Safety Administration 400 7th Street SW. Washington, DC 20590			10. SPONSORING/MONITORING AGENCY REPORT NUMBER DOT HS 810 580
11. SUPPLEMENTARY NOTES			
12a. DISTRIBUTION/AVAILABILITY STATEMENT This document is available to the public through the National Technical Information Service, Springfield, Virginia 22161.			12b. DISTRIBUTION CODE

13. ABSTRACT (Maximum 200 words)

This report presents results from an analysis of about 47,000 red light violation records collected from 11 intersections in the City of Sacramento, California, by red light photo enforcement cameras between May 1999 and June 2003. The goal of this analysis is to understand the correlation between red light violations and various driver, intersection, and environmental factors. Descriptive statistics suggest that younger drivers under 30 years of age are more likely to run red lights than drivers in other age groups. About 56 percent of the violators were traveling at or below the posted speed limit. Moreover, 94 percent of the violations occurred within 2 seconds after the onset of red light, and only 3 percent of the violations were recorded 5 seconds after the onset of red light. Approximately 4 percent of the violators were repeat offenders. Logistic regression modeling shows that the predicted odds of a younger driver running a red light at speeds greater than the speed limit is about 1.5 times the odds of a middle-aged driver. In addition, older drivers have a higher probability of running a red light when the elapsed time since the onset of red light is more than 2 seconds compared to younger drivers. Finally, red light violations rates are estimated between 6 and 29 violations per 100,000 intersection-crossing vehicles.

14. SUBJECT TERMS Red light violations, signalized intersections, Intelligent Vehicle Initiative, red light photo enforcement cameras, cooperative signal violation warning system			15. NUMBER OF PAGES 70
			16. PRICE CODE
17. SECURITY CLASSIFICATION OF REPORT Unclassified	18. SECURITY CLASSIFICATION OF THIS PAGE Unclassified	19. SECURITY CLASSIFICATION OF ABSTRACT Unclassified	20. LIMITATION OF ABSTRACT

NSN 7540-01-280-5500

Standard Form 298 (Rev. 2-89)
Prescribed by ANSI Std. 239-18
298-102

PREFACE

The National Highway Traffic Safety Administration (NHTSA), in conjunction with the Research and Innovative Technology Administration's Volpe National Transportation Systems Center (Volpe Center), conducted a preparatory analysis for a potential field operational test of a vehicle-intersection cooperative signal violation warning system that addresses crossing path crashes at signalized intersections. This analysis supports the Intelligent Vehicle Initiative (IVI) of the U.S. Department of Transportation. The IVI accelerates the development and deployment of vehicle-based and vehicle-infrastructure cooperative crash countermeasures using advanced technologies over several problem areas: rear-end, roadway departure, lane change, crossing paths, driver impairment, reduced visibility, vehicle instability, pedestrian, and pedalcyclist crashes.

This report presents the results obtained from the data analysis of about 47,000 red light violation records collected by the City of Sacramento, California, using red light photo enforcement cameras at 11 intersections during the time period between May 1999 and June 2003.

The authors of this report are C. Y. David Yang and Wassim G. Najm of the Volpe Center.

The authors acknowledge the technical contribution of Dr. David L. Smith and Kerrin Bressant of NHTSA. Special thanks are extended to Robert Ferlis of the Federal Highway Administration and Matthew T. Schmitz at the California Division of the Federal Highway Administration for recommending and contacting the City of Sacramento regarding the red light violation data used for this study. The authors also want to thank Police Chief Albert Najera and Sergeant Eric Poerio of the Sacramento Police Department for their willingness to share the red light violation data with the Volpe Center. Moreover, Angie Louie Fong and her staff from the Sacramento Department of Public Works' Traffic Engineering Services were very helpful in providing useful traffic and infrastructure information regarding the City of Sacramento's signalized intersections. Finally, the authors would like to acknowledge John Flynn, Lon B. Ecklund, and their staff at Affiliated Computer Services, Inc. (ACS) for their diligent efforts in organizing and sanitizing Sacramento's red light violation records so they could be used for this study. (Note: Sacramento Police Department hired ACS to handle and process its red light violation data.)

METRIC/ENGLISH CONVERSION FACTORS

ENGLISH TO METRIC | METRIC TO ENGLISH

LENGTH (APPROXIMATE)

ENGLISH TO METRIC	METRIC TO ENGLISH
1 inch (in) = 2.5 centimeters (cm)	1 millimeter (mm) = 0.04 inch (in)
1 foot (ft) = 30 centimeters (cm)	1 centimeter (cm) = 0.4 inch (in)
1 yard (yd) = 0.9 meter (m)	1 meter (m) = 3.3 feet (ft)
1 mile (mi) = 1.6 kilometers (km)	1 meter (m) = 1.1 yards (yd)
	1 kilometer (km) = 0.6 mile (mi)

AREA (APPROXIMATE)

ENGLISH TO METRIC	METRIC TO ENGLISH
1 square inch (sq in, in^2) = 6.5 square centimeters (cm^2)	1 square centimeter (cm^2) = 0.16 square inch (sq in, in^2)
1 square foot (sq ft, ft^2) = 0.09 square meter (m^2)	1 square meter (m^2) = 1.2 square yards (sq yd, yd^2)
1 square yard (sq yd, yd^2) = 0.8 square meter (m^2)	1 square kilometer (km^2) = 0.4 square mile (sq mi, mi^2)
1 square mile (sq mi, mi^2) = 2.6 square kilometers (km^2)	10,000 square meters (m^2) = 1 hectare (ha) = 2.5 acres
1 acre = 0.4 hectare (he) = 4,000 square meters (m^2)	

MASS - WEIGHT (APPROXIMATE)

ENGLISH TO METRIC	METRIC TO ENGLISH
1 ounce (oz) = 28 grams (gm)	1 gram (gm) = 0.036 ounce (oz)
1 pound (lb) = 0.45 kilogram (kg)	1 kilogram (kg) = 2.2 pounds (lb)
1 short ton = 2,000 pounds (lb) = 0.9 tonne (t)	1 tonne (t) = 1,000 kilograms (kg)
	= 1.1 short tons

VOLUME (APPROXIMATE)

ENGLISH TO METRIC	METRIC TO ENGLISH
1 teaspoon (tsp) = 5 milliliters (ml)	1 milliliter (ml) = 0.03 fluid ounce (fl oz)
1 tablespoon (tbsp) = 15 milliliters (ml)	1 liter (l) = 2.1 pints (pt)
1 fluid ounce (fl oz) = 30 milliliters (ml)	1 liter (l) = 1.06 quarts (qt)
1 cup (c) = 0.24 liter (l)	1 liter (l) = 0.26 gallon (gal)
1 pint (pt) = 0.47 liter (l)	
1 quart (qt) = 0.96 liter (l)	
1 gallon (gal) = 3.8 liters (l)	
1 cubic foot (cu ft, ft^3) = 0.03 cubic meter (m^3)	1 cubic meter (m^3) = 36 cubic feet (cu ft, ft^3)
1 cubic yard (cu yd, yd^3) = 0.76 cubic meter (m^3)	1 cubic meter (m^3) = 1.3 cubic yards (cu yd, yd^3)

TEMPERATURE (EXACT)

ENGLISH TO METRIC	METRIC TO ENGLISH
[(x-32)(5/9)] F = y C	[(9/5) y + 32] C = x F

QUICK INCH - CENTIMETER LENGTH CONVERSION

QUICK FAHRENHEIT - CELSIUS TEMPERATURE CONVERSION

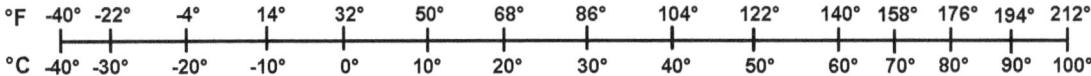

For more exact and or other conversion factors, see NIST Miscellaneous Publication 286, Units of Weights and Measures. Price $2.50 SD Catalog No. C13 10286.

Updated 6/17/98

TABLE OF CONTENTS

EXECUTIVE SUMMARY vii

1. INTRODUCTION 1
   1.1. Background 1
   1.2. Report Outline 2

2. LITERATURE REVIEW ON RED LIGHT VIOLATION 3
   2.1. Definitions of Red Light Violation 3
   2.2. Effects of Driver, Intersection, and Environment on Red Light Violations 5
   2.3. Red Light Violation Countermeasures 8

3. STATISTICAL DESCRIPTION OF RED LIGHT VIOLATIONS 9
   3.1. Overview of Sacramento's Red Light Running Program 9
   3.2. Picture Processing Procedures 10
   3.3. Descriptive Statistics 11
   3.3.1. General Information on Red Light Violation Records 11
   3.3.2. Information on RLPEC-Equipped Signalized Intersections 13
   3.3.3. Frequency Distributions of Selected Variables 14
   3.4. Estimates of Red Light Violation and Crash Rates 25

4. STATISTICAL MODELING OF RED LIGHT VIOLATION DATA 30
   4.1. Logistic Modeling Method 30
   4.2. Factors with Significant Influence on Violators' Vehicle Speed 31
   4.3. Factors with Significant Influence on Time Elapsed since Red Light Onset . 35

5. CONCLUDING REMARKS 39
   5.1. List of Major Findings 39
   5.2. Implications of Findings 40

6. REFERENCES 41

Appendix A. Photos of City of Sacramento's 11 RLPEC-Equipped
   Intersections 44

Appendix B. Photos of Red Light Photo Enforcment Camera and Warning
   Sign 50

Appendix C. GIS Map of a Selected RLPEC-Equipped Intersection in
   Sacramento 51

Appendix D. Distributions of Red Light Violations by Speed at Individual
   11 RLPEC-Equipped Intersections 52

LIST OF FIGURES

Figure 1. Location of Sacramento's 11 RLPEC-Equipped Intersections 12
Figure 2. Normalized Red Light Violation Values by Age Group 16
Figure 3. Distribution of Red Light Violations by Time of Day 17
Figure 4. Distribution of Red Light Violation Records by Vehicle Year 18
Figure 5. Distribution of Red Light Violation Records by Vehicle Speed 19
Figure 6. Percentage Cumulative Distribution of Red Light Violation Records by Vehicle Speed 19
Figure 7. Percentage Distribution of Violation Records by [Vehicle Speed – PSL] 21
Figure 8. Percentage Cumulative Distribution of Records by [Vehicle Speed – PSL] ... 21
Figure 9. Distribution of Violation Records by Elapsed Time Since Red Light Onset (Vertical Axis in Logarithmic Scale) 22
Figure 10. Percentage Distribution of Violation Records by Time Elapsed Since Red Light Onset 22
Figure 11. Percentage Cumulative Distribution of Violations by Time Elapsed Since Red Light Onset 23
Figure 12. Picture of Fair Oaks Boulevard and Howe Avenue Intersection 44
Figure 13. Picture of El Camino Avenue and Evergreen Street Intersection 44
Figure 14. Picture of Arden Way and Exposition Boulevard Intersection 45
Figure 15. Picture of Mack Road and La Mancha Way/Valley Hi Drive Intersection.... 45
Figure 16. Picture of Mack Road and Center Parkway Intersection 46
Figure 17. Picture of 30th Street and Capitol Avenue Intersection 46
Figure 18. Picture of J Street and Alhambra Boulevard Intersection 47
Figure 19. Picture of Broadway and 21st Street Intersection 47
Figure 20. Picture of W Street and 16th Street [Highway 50 Exit at 16th Street] Intersection... 48
Figure 21. Picture of Howe Avenue and College Town Drive Intersection 48
Figure 22. Picture of Power Inn Road and Folsom Boulevard Intersection 49
Figure 23. Picture of Red Light Photo Enforcment Camera in the City of Sacramento.. 50
Figure 24. Picture of Red Light Photo Enforcment Camera Warning Sign in the City of Sacramento... 50
Figure 25. GIS Mapping of Fair Oaks Boulevard and Howe Avenue Intersection 51
Figure 26. Frequency Distribution of Red Light Violations by Speed at Fair Oaks Boulevard and Howe Avenue Intersection 52
Figure 27. Frequency Distribution of Red Light Violations by Speed at El Camino Avenue and Evergreen Street Intersection... 53
Figure 28. Frequency Distribution of Red Light Violations by Speed at Arden Way and Exposition Boulevard Intersection... 53
Figure 29. Frequency Distribution of Red Light Violations by Speed at Mack Road and La Mancha Way/Valley Hi Drive Intersection 54
Figure 30. Frequency Distribution of Red Light Violations by Speed at Mack Road and Center Parkway Intersection 54
Figure 31. Frequency Distribution of Red Light Violations by Speed at 30th Street and Capitol Avenue Intersection 55

Figure 32. Frequency Distribution of Red Light Violations by Speed at J Street and Alhambra Boulevard Intersection 55

Figure 33. Frequency Distribution of Red Light Violations by Speed at Broadway and 21st Street Intersection.. 56

Figure 34. Frequency Distribution of Red Light Violations by Speed at W Street and 16th Street [Highway 50 Exit at 16th Street] Intersection 56

Figure 35. Frequency Distribution of Red Light Violations by Speed at Howe Avenue and College Town Drive Intersection 57

Figure 36. Frequency Distribution of Red Light Violations by Speed at Power Inn Road and Folsom Boulevard Intersection 57

LIST OF TABLES

Table 1. Definitions of Red Light Violation by Different Studies/Jurisdictions 4
Table 2. Effects of Driver, Intersection, and Environment on Red Light Violations 5
Table 3. Red Light Violation Rates Reported by Previous Studies 7
Table 4. Suggested Red Light Violation Countermeasures 8
Table 5. Distribution of Red Light Violation Records and Their Collection Period 13
Table 6. Traffic and Infrastructure Characteristics of the 11 RLPEC-Equipped Intersections 15
Table 7. Distribution of Red Light Violation Records by Age Group 16
Table 8. Distribution of Red Light Violation Records by Gender 17
Table 9. Distribution of Violation Records by Vehicle Speed in Comparison to the Posted Speed Limit 20
Table 10. Distribution of Red Light Violation Records by Selected Categories of Time Elapsed Since Red Light Onset 23
Table 11. Percentage Cumulative Distribution of Violations by Elapsed Time Since Red Light Onset for Various Vehicle Speeds 24
Table 12. Percentage Cumulative Distribution of Violations by Elapsed Time Since Red Light Onset for Various Delta Speeds 24
Table 13. Statistics on Repeat Red Light Offenders 25
Table 14. Estimates of Red Light Violation Rates from 11 RLPEC-Equipped Intersections 27
Table 15. Collision Rates from Sacramento's RLPEC-Equipped Intersections 29
Table 16. Dependent and Explanatory Variables Used in Logistic Regression Analysis 32
Table 17. Estimation Results for the Binary Logit Model with "Speed2" as Dependent Variable 33
Table 18. Estimation Results from Binary Logit Model with "RedEla2" as Dependent Variable 37

LIST OF ACRONYMS

ACS	Affiliated Computer Services
ADT	Average Daily Traffic
CSVWS	Cooperative Signal Violation Warning System
FOT	Field Operational Test
GIS	Geographic Information System
ITE	Institute of Transportation Engineers
IVI	Intelligent Vehicle Initiative
LD	Licensed Driver
MUTCD	Manual on Uniform Traffic Control Devices
MVMT	Million Vehicle Miles Traveled
NHTSA	National Highway Traffic Safety Administration
PSL	Posted Speed Limit
RLPEC	Red Light Photo Enforcement Camera
RLV	Red Light Violation

EXECUTIVE SUMMARY

The goal of the Intelligent Vehicle Initiative (IVI) of the U.S. Department of Transportation is to accelerate the development and deployment of vehicle-based and vehicle-infrastructure cooperative crash countermeasures using advanced technologies over several problem areas: rear-end, roadway departure, lane change, crossing paths, driver impairment, reduced visibility, vehicle instability, pedestrian, and pedalcyclist crashes. As part of the crossing path crash problem area, the National Highway Traffic Safety Administration has tasked the Volpe National Transportation Systems Center (Volpe Center) to conduct preparatory analysis for a potential field operational test of a vehicle-intersection cooperative signal violation warning system (CSVWS). It is envisioned that such system will provide advisory/warning messages to the driver of a moving vehicle who is about to run a red light because of failure to:

- Recognize the presence and status of the traffic signal, or
- Judge the adequate time to safely clear a signalized intersection.

To develop, test, and design an effective CSVWS that can prevent red light violation, it is necessary to identify the causal factors and circumstances of red light violations. It is also important to understand the correlation between red light violations and various driver, intersection, and environmental factors.

With assistance from the Police and Public Works Departments in the City of Sacramento, California, the Volpe Center has received data from four years of red light violation records that were gathered by red light photo enforcement cameras (RLPECs) at 11 signalized intersections in the City of Sacramento. Volpe Center staff analyzed the red light violation data to identify factors with strong correlation to red light running behavior. Findings from this study are important to the estimation of safety benefits, and development of performance specifications and objective test procedures for the CSVWS.

RLPECs at the City of Sacramento's signalized intersections are activated to photograph potential violators when the following conditions are met:

- Vehicle enters the intersection after the signal light had been red for a minimum elapsed time of 0.2 second, and
- The measured speed of violating vehicle is more than 15 mph (on straight-through lane) or 13 mph (on left-turn lane).

Several pictures are taken to capture the vehicle crossing the stop bar after the onset of a red signal, license plate number of the violating vehicle, and the person who is driving the violating vehicle. A review of these pictures is then conducted by the police to decide on whether or not a citation is warranted. No violation citation is issued if:

- Pictures do not clearly show that the vehicle crossed the stop bar after the signal turned red.

- Registered owner of the violating vehicle does not match the driver who ran the red light.

Based on one estimate, only 35 percent of the pictures taken by Sacramento's RLPECs turn into violation citations. Red light violation records used in this analysis only include the actual violation citations issued by the Sacramento Police Department. Data from a total of 46,997 violation records (May 1999 to June 2003) were provided to the Volpe Center. The following variables are included in this data set:

- Intersection (and intersection code) where the violation occurred
- Date when the violation occurred
- Time when the violation occurred
- Age of the violator
- Gender of the violator
- Car (i.e., vehicle make) driven by the violator
- Model year of the vehicle driven by the violator
- Vehicle speed (i.e., measured speed) at the time of the violation
- Elapsed time from the onset of the red signal until the time of the violation

Descriptive statistics from the analysis of the red light violation data suggest that younger drivers under 30 years of age are more likely to run the red light than drivers in other age groups. In addition, most red light violations occur during the daytime (i.e., 7 a.m. to 7 p.m.) with highest counts of red light violations during the period from 2:00 p.m. to 2:59 p.m. The three most frequent vehicle speeds at the time of the violation were: 18 mph (1,762 records), 17 mph (1,715 records), and 19 mph (1,711 records). The average red light violation speed was 31.6 mph. It should be noted that the highest posted speed limit among these 11 RLPEC-equipped intersections is 45 mph. About 18 percent of the violators ran the red light at speeds higher than 45 mph. Moreover, about 56 percent of the violators were traveling at or below the posted speed limit (i.e., not speeding). The City of Sacramento's RLPECs have captured drivers crossing the intersection from 0.2 second (6,381 records) to more than 30 seconds (434 records) after the onset of the red light. Approximately 94 percent of the red light violations occurred within 2 seconds after the onset of the red light, and only 3 percent of the violations were recorded after 5 seconds from the onset of the red light. As for repeat offenders, about 4 percent of the violators had more than one red light violation.

Logistic regression modeling indicates that as the age of the red light violator increases, the probability of running the red light at vehicle speeds greater than the posted speed limit (i.e., speeding) decreases. The predicted odds of a younger driver running a red light while speeding is about 1.5 times the odds of a middle-aged driver doing so. Moreover, logistic regression modeling also shows that drivers who run red lights between 6 a.m. and 7 p.m. have a lower probability of speeding than violators in the time period between 7 p.m. and 6 a.m. Furthermore, red light violators at intersections with heavy traffic volumes have a lower probability of speeding.

Another logistic regression model shows that older drivers have a higher probability of running a red light when the elapsed time since the onset of red light is more than 2 seconds compared to younger drivers. Also, drivers who run red lights between 6 a.m. and 7 p.m. have a lower probability of crossing intersections when the elapsed time from the onset of red light is more than 2 seconds compared to motorists who run red lights between 7 p.m. and 6 a.m. Another key finding from the logistic regression model is that motorists who run a red light at the intersection with the highest clearance interval (i.e., yellow time and all-red time) are more likely to drive through the intersection when the elapsed time since the onset of red light is more than 2 seconds.

Finally, red light violation rates (calculated from issued violation citations) for the 11 RLPEC-equipped intersections are estimated between 6 and 29 violations per 100,000 crossing vehicles. In comparison to red light violation rates reported by other studies, these estimated red light violation rates are quite low. This is probably due to the fact that these estimated rates were based on actual police citations. It should be noted that only 35 percent of the photos taken by the City of Sacramento's RLPECs become red light violation citations. The results of this study have demonstrated that drivers in different age groups exhibit diverse behavior when approaching signalized intersections. Hence, the experimental design for the CSVWS field operational test must examine the influence of driver age. The logistic regression models developed in this study suggest that several variations of the CSVWS warning algorithm and warning messages might be necessary for different time periods throughout the day. At certain time period(s) when drivers are susceptible to speeding through intersections or entering intersections late when the light changes, CSVWS warnings need to be issued earlier and warning messages need to be decisive to effectively encourage more drivers to stop for the red light.

1. INTRODUCTION

1.1. Background

A total of 9,951 vehicles were involved in fatal crashes at traffic signals in 1999 and 2000 based on the Fatality Analysis Reporting System crash database, with 20 percent of these vehicles failing to obey the signals (Campbell et al., 2004). From 1992 to 1998, an estimated 1.5 million people were injured in crashes related to red light violation (Insurance Institute for Highway Safety, 2000). Intersections are among the most dangerous locations on U.S. roadways, and red light running is a major transportation safety challenge at signalized intersections.

The Intelligent Vehicle Initiative (IVI) of the U.S. Department of Transportation (DOT) was established in 1998 to address a number of crash types by accelerating the development, availability, and use of driving assistance and control intervention systems. The IVI emphasizes the significant and continuing role of drivers in roadway safety, and aims at helping drivers process information, make decisions, and operate vehicles more safely.

As part of the crossing path crash problem area, the National Highway Traffic Safety Administration (NHTSA) has tasked the Volpe National Transportation Systems Center (Volpe Center) to conduct preparatory analysis for a potential field operational test (FOT) of a vehicle-intersection cooperative signal violation warning system (CSVWS). It is envisioned that such a system will provide advisory/warning messages to the driver of a moving vehicle who is about to run a red light because of failure to:

- Recognize the presence and status of the traffic signal, or
- Judge the adequate time to safely clear a signalized intersection.

When warning messages are issued, the driver decides on the appropriate action to take in response to the warnings. Conceptually, the CSVWS will have an infrastructure-based component and a vehicle-based component. The infrastructure-based component establishes a communication channel between the operational units of a signalized intersection (e.g., controller and detection system) and a moving vehicle. The vehicle-based component of the CSVWS acts on the information provided by the intersection and vehicle sensors to detect the potential of a red light running activity and provide suitable warning messages to the driver.

To develop, test, and design an effective CSVWS that can prevent red light violations, it is necessary to identify the causal factors and circumstances of red light violations. Specifically, the correlation between red light violations and various driver, intersection, and environmental factors needs to be studied so as to address the following questions:

- Are there certain types of drivers with higher red light running tendency?
- Are there some intersection characteristics that contribute to higher red light violation rates?
- Are there certain environmental factors that lead to higher red light violation rates?

With assistance from the Police and Public Works Departments in the City of Sacramento, California, the Volpe Center has received data from four years of red light violation records gathered by the red light photo enforcement cameras (RLPECs) at 11 signalized intersections in Sacramento. Volpe Center staff analyzed the data to identify factors with strong correlation to red light running behavior. Significant factors determined by this research effort are important to the estimation of safety benefits, and development of performance specifications and objective test procedures for the CSVWS. In addition, red light violation rates from these 11 RLPEC-equipped intersections are estimated and compared to violation rates reported by other researchers.

1.2. Report Outline

This report consists of five major sections. The first section presents the objectives of this study and related background information. The second section summarizes red light definitions adopted by other researchers, and lists the factors that showed strong correlation to red light violation as well as various red light violation rates reported by previous researchers. The third section introduces the red light violation data gathered by the RLPECs in the City of Sacramento, California. Moreover, Sacramento's Red Light Running Program is described along with the procedures used to process red light violation citations. The third section also presents some descriptive statistics about red light violations. Furthermore, red light violation and crash rates at RLPEC-equipped intersections in Sacramento are also estimated. The fourth section delineates significant factors with strong correlation to red light violation, which were determined using statistical methods. Finally, the fifth section of this report summarizes and discusses the implications of key findings from the analysis of Sacramento's red light violation data.

2. LITERATURE REVIEW ON RED LIGHT VIOLATION

This section summarizes results of previous studies on red light violation, which help to address the following questions:

1. What constitutes a red light violation?
2. What driver, intersection, and roadway characteristics are prevalent to red light violations?
3. What are some of the strategies recommended for the reduction of red light violations?

2.1. Definitions of Red Light Violation

The Manual on Uniform Traffic Control Devices (December 2000), or MUTCD, defines traffic signal indications (i.e., green, yellow, and red signals) as follows:

- Steady green signal indicates that vehicle "may proceed straight or turn right or left (at an intersection) except as such movement is modified by lane-use signs, turn prohibition signs, lane markings, or roadway design."
- Steady yellow signal indicates that vehicle "is thereby warned that the related green movement is being terminated or that a red signal indication will be exhibited immediately thereafter when vehicular traffic shall not enter the intersection".
- Steady red signal indicates that vehicle "shall stop at a clearly marked stop line, but if there is no stop line, traffic shall stop before entering the crosswalk on the near side of the intersection; or if there is no crosswalk, then before entering the intersection, and shall remain stopped until a signal indication to proceed is shown."

Based on the definitions listed above, drivers should be prepared to stop when they see the yellow signal and stop completely when the signal changes to red. Unfortunately, many drivers fail to observe and comply with traffic signal changes. Consequently, red light violation is one of the major causes for crossing path crashes at signalized intersections. Research studies have reported various factors for red light violations such as timing of the traffic signal, driving environment, and driver characteristics. The probability of crossing path crashes occurring at a signalized intersection increases as the rate of red light violation increases. A clear definition of a red light violation by the MUTCD is "… when a driver fails to stop at the presence of a steady red signal indication." However, the interpretation for red light violation changes among the various studies and jurisdictions as presented in Table 1.

As seen in Table 1, there is no consistent definition of red light violation. Many factors such as engineering considerations, environmental variables, and driver demographics could all have an effect on a local jurisdiction's policy for red light violation. For instance, the interpretation of a red light violation in a large city with hilly streets and a serious traffic congestion problem would probably be lax compared to a rural town with level streets and rare traffic congestion.

The City of Sacramento's RLPECs are activated to photograph potential violators when the following conditions are met:

- Vehicle enters the intersection after the signal light had been red for a minimum elapsed time of 0.2 second, and
- The measured speed of violating vehicle is more than 15 mph (on straight-through lane) or 13 mph (on left-turn lane).

Table 1. Definitions of Red Light Violation by Different Studies/Jurisdictions

Definition	Reference	Study Background
Vehicle enters the intersection after the signal light had been red for more than 0.2 second and the measured speed of vehicle is ≥ 18 mph for roads with speed limits of 45 mph or less and 20 mph threshold for roadways with higher posted speed limits.	Ruby and Hobeika, 2003	This study assesses the Red Light Running camera enforcement program in Fairfax County, Virginia, where 10 cameras were installed at high commuter traffic intersections.
Vehicle moves across the stop bar after the signal phase changed from yellow to red.	Lum and Wong, 2003	This is a before-and-after study that evaluated the impacts of installing and operating red light cameras at two "T" and one "X" signalized intersections.
	Schattler et al., 2002	Video cameras collected data from 3 intersections in Oakland County, Michigan, for a before-and-after evaluation of the impacts of clearance intervals on red light violation rates.
	Kamyab et al., 2002; Kamyab et al., December 2000	The Center for Transportation Research and Education at Iowa State University collected red light running data at 12 intersections from 7 Iowa communities.
Vehicle enters the intersection after the signal light had been red for a minimum elapsed time of 0.4 second and the measured speed of vehicle is at least 15 mph.	Retting et al., 1999a	Baseline red light violation data were collected at 9 sites equipped with red light violation cameras in a red light camera enforcement study conducted in Oxnard, California.
	Retting et al., 1999b	Baseline red light violation data were collected at 5 sites in Fairfax, Virginia, which have histories of crashes involving red light running.
Vehicle enters the intersection any time after onset of the red signal and is traveling at least 15 mph.	Retting et al., 1998	Red light cameras collected data at two busy intersections in Arlington, Virginia.
Vehicle enters the intersection ≥ 0.5 second after the onset of a red traffic signal.	Retting and Williams, 1996	An automated camera and trained observers collected red light running data at an intersection with an eight-lane east/west principal arterial and a four-lane north/south collector street in Arlington County, Virginia.

2.2. Effects of Driver, Intersection, and Environment on Red Light Violations

Table 2 presents various variables that showed some effects on drivers' red light violation behavior. Table 3 lists red light violation rates reported in previous studies.

Table 2. Effects of Driver, Intersection, and Environment on Red Light Violations

Element	Variable	Key Finding	Reference
Driver	*Age*	The older age groups accounted for a relatively small portion of red light running crashes compared to the young age group.	Kraus and Quiroga, 2004
		Younger drivers between the ages of 18 to 25 years old are more likely to run red lights compared to other age groups.	Porter and Berry, 2001
		Red light runners tend to be drivers under 30 years old.	Retting et al., 1999; Retting and Williams, 1996
	Gender	Red light runners are more likely than non-runners to be male.	Retting et al., 1999
	Occupancy	Drivers have a higher probability of running red lights when driving alone compared to when passengers are in their vehicles.	Porter and Berry, 2001
	Safety Belt	Red light runners are less likely to wear safety belts.	Porter and England, 2000; Retting and Williams, 1996
	Driving Record	Red light runners are more likely than non-runners to be driving with suspended or revoked driver's licenses.	Retting et al., 1999
		Drivers with poor driving records and driving smaller and older cars have a higher tendency to run red lights.	Retting and Williams, 1996
Intersection	*Signal Timing*	The frequency of red light running increases when the yellow interval is less than 3.5 seconds.	Brewer et al., 2002
		Longer yellow intervals will cause drivers to enter intersection later and lengthening the all-red intervals caters to red light violators.	Eccles and McGee, July 2001
	Stopping Distance	Probability of a vehicle stopping for a traffic light decreases as its distance from the intersection decreases.	Chang et al., 1985
	Approach Speed	Probability of a driver stopping for a traffic light decreases as the approach speed to the intersection increases.	Chang et al., 1985
	Grade	Probability of a driver stopping for a traffic light increases as the grade of the approaching intersection increases (i.e., roadway becomes "steeper").	Chang et al., 1985
	Intersection Width	Drivers tend to stop for traffic lights more at wider intersections than at narrower intersections.	Chang et al., 1985

Table 2. Effects of Driver, Intersection, and Environment on Red Light Violations (Cont.)

Element	Variable	Key Finding	Reference
Traffic & Environment	*Approach Volume*	Higher red light running rates were observed in cities with wider intersections and higher traffic volumes.	Porter and England, 2000
		The red light running frequency increases as the approach traffic volume at intersections increases.	Brewer et al., 2002
	Time of Day	Higher red light violations occur during the time period of 3:00 PM to 5:00 PM.	Kamyab, et al., 2002; Kamyab, et al., December 2000
		The average red light violations are higher during AM and PM peak hours compared to other times of the day.	Retting et al., 1998
	Day of Week	There are more red light violations on weekdays compared to weekends.	Lum and Wong, 2003; Kamyab, et al., 2002; Kamyab, et al., December 2000; Retting et al., 1998
	Weather	The influence of rainfall on red light running behavior is insignificant.	Retting et al., 1998

Major observations made from the information presented in Table 2 and Table 3 include:

- According to Table 2, factors ranging from age and gender of the driver, signal timing, and approach speed of the intersection, to time of day could all affect red light running behavior. Factors with strong influence on red light violation could be region-specific or location-specific.
- A wide range of violation rates has been reported by different studies. Factors such as definitions of red light violation used in studies, methods used for data collection, and locations of studies could all contribute to variations in violation rates. Information presented in Table 3 confirmed that red light violation is a common problem in many communities; however, the effects of red light violation from one location to the next can vary considerably.
- Studies that examined the before-and-after effect of implementing red light photo enforcement cameras suggest that such systems have a great potential to effectively reduce red light violations. However, some researchers pointed out that the full consequences of red light cameras are still being studied. For example, McGee (2002) indicated that there is still no conclusive answer yet about the effect of red light photo enforcement cameras on crash rates at signalized intersections due to the lack of comprehensive and statistically rigorous study designs.

Table 3. Red Light Violation Rates Reported by Previous Studies

Reference	Rate Before Project Implementation	Rate After Project Implementation
Lum and Wong, 2003	Average weekday red light violations ranging from 16.0 to 111.8 per day at two "T" intersections before implementation of red light cameras.	Weekday red light violations reduced to 13.4 to 58.6 per day at two "T" intersections after installation of red light cameras.
Rudy and Hobeika, 2003	10 intersections with various red light violation rates ranging from 2.00 violations to 11.0 violations per 10,000 vehicles.	3 months after installation of red light running cameras, violation rates at these intersections were reduced to between 0.17 violation and 7.0 violations per 10,000 vehicles.
Brewer et al., 2002	An overall average of 4.1 red light runners per 1,000 vehicles.	N/A
Schattler et al., 2002	3 intersections with various rates ranging from 0 violations per hour to an average of 10.2 violations per hour.	Violation rates ranging from 0 violations per hour to an average of 4.6 violations per hour after the implementation of new clearance intervals per ITE guidelines.
Fakhry and Salaita, 2002	An average of 1.3 red light violations per 1,000 vehicles (manual observation).	N/A
Kamyab, et al., 2002; Kamyab, et al., December 2000	13 intersections with various violation rates ranging from 0.45 violation per 1,000 entering vehicles to 38.50 violations per 1,000 vehicles.	N/A
Porter and England, 2000	At least one red light runner in 1,798 out of the 5,112 observed light cycles (35.2%), roughly 10 red light violations per observed hour.	N/A
Retting et al., 1999a	12.9 violations per 10,000 vehicles at red light camera sites and 16.0 violations per 10,000 vehicles at non-camera sites.	7.7 violations per 10,000 vehicles at red light camera sites and 8.0 violations per 10,000 vehicles at non-camera sites 4 months after the implementation of red light cameras.
Retting et al., 1999b	36.3 violations per 10,000 vehicles at red light camera sites and 37.8 violations per 10,000 vehicles at non-camera sites.	20.4 violations per 10,000 vehicles at red light camera sites and 25.0 violations per 10,000 vehicles at non-camera sites 1 year after the installation of red light cameras.
Retting et al., 1998	6,171 violations observed during 1,176 hours on Site 1 (5.2 runners/hour) and 1,950 violations observed during 1,518 hours on Site 2 (1.3 runners/hour).	N/A
Retting and Williams, 1996	462 red light violations out of 1,373 observations made over 234 hours of data collection, about 2 red light runners per hour of data collection.	N/A

ITE ≡ Institute of Transportation Engineers

2.3. Red Light Violation Countermeasures

Table 4 lists countermeasures suggested by researchers to address red light violations. As alternatives to photo enforcement cameras, some countermeasures focus on improving the traffic control system at intersections and the physical layout of the intersections. In certain situations, the use of "non-enforcement" countermeasures would be sufficient to effectively reduce red light violations at intersections.

Table 4. Suggested Red Light Violation Countermeasures

Countermeasure	Reference
Use automated red light photo enforcement cameras.	Rudy and Hobeika, 2003; Retting and Kyrychenko, 2002; BMI, December 2001; Retting and Kyrychenko, April 2001; Retting et al., 1999; Retting et al., 1998; Retting and Williams, 1996
Adjust the timing of traffic signals with adequate yellow clearance interval time/in accordance with recommendations made by ITE.	Brewer et al., 2002; Milazzo et al., 2002; Schattler et al., 2002; BMI, December 2001; Milazzo et al., June 2001; Retting et al., September 2000; Retting et al., 1999; Retting et al., 1998; Retting and Greene, 1997; Retting and Williams, 1996; Retting et al., 1995; Zador et al., 1985
Increase signal and sign visibility.	Brewer et al., 2002; BMI, December 2001; Retting et al., 1995
Improve signal coordination among intersection groups.	Brewer et al., 2002
Use protected left-turn phases.	BMI, December 2001
Provide all-red intervals at intersections.	Retting et al., 1995
Increase sight distance.	Retting et al., 1995

ITE ≡ Institute of Transportation Engineers

This section of the report provides a summary of literature on topics related to red light violation. This information forms the "baseline" knowledge in understanding red light violation definition and the effects of driver, intersection, and environment on red light violation. Sections 3 and 4 of this report will present and discuss results from the analysis of red light violation data gathered from the City of Sacramento's 11 RLPEC-equipped intersections.

3. STATISTICAL DESCRIPTION OF RED LIGHT VIOLATIONS

This section statistically describes the red light violation data obtained from the City of Sacramento's Red Light Running Program. It is important to note that the violation records given to the Volpe Center contained no personal information of red light offenders. To protect the privacy of red light offenders, information such as offenders' names, mailing addresses, and vehicle license plate numbers were all removed from the data set before its transfer to the Volpe Center.

3.1. Overview of Sacramento's Red Light Running Program

The Sacramento Police Department reported about 5,500 vehicle collisions at intersections in the City of Sacramento in 1998. Approximately 13 percent of these collisions were caused by running the red light, resulting in 494 injuries and a financial cost of more than $15 million to the local economy.[1] Upon reviewing these statistics, Sacramento's City Council decided to implement a new program that aims to enhance safety and modify driver behavior at signalized intersections.

After going through an evaluation process, the City of Sacramento decided to deploy RLPECs at selected signalized intersections to capture vehicles/drivers deliberately running the red light. On December 15, 1998, Sacramento's City Council awarded a service contract to a private company to install cameras at 11 signalized intersections throughout the city. The selection of these intersections was based on three criteria:

- Number of intersection collisions caused by red light violations
- Total intersection traffic volumes
- Inputs from the police department and the community

RLPECs were installed at one or more approaches (or directions) of the following 11 intersections in the City of Sacramento:

1. Fair Oaks Boulevard and Howe Avenue
2. El Camino Avenue and Evergreen Street
3. Arden Way and Exposition Boulevard
4. Mack Road and La Mancha Way/Valley Hi Drive
5. Mack Road and Center Parkway
6. 30th Street and Capitol Avenue
7. J Street and Alhambra Boulevard
8. Broadway and 21st Street
9. W Street and 16th Street
10. Howe Avenue and College Town Drive
11. Power Inn Road and Folsom Boulevard

---

[1] Data presented above can be found at
http://www.cityofsacramento.org/transportation/engineering/trafficredlight.html

The intersection of Fair Oaks Boulevard and Howe Avenue was identified by a 2001 study as one of the 10 most dangerous intersections in the United States (American City & County, 2001). Pictures for the 11 intersections listed above are provided in Appendix A.

The City of Sacramento began to issue red light violation warning notices to offenders for a period of 30 days, as required by the California Vehicle Code, on May 26, 1999. On June 26, 1999, the City began to issue red light violation citations to offenders. Sacramento's Red Light Running Program is still ongoing. Appendix B presents pictures of a RLPEC and a warning sign notifying drivers that they are approaching an intersection equipped with enforcement cameras.

3.2. Picture Processing Procedures

This subsection describes the procedures by the Sacramento Police Department to process RLPEC pictures on potential red light offenders. As indicated earlier, RLPECs at the City of Sacramento's signalized intersections are activated to photograph potential violators when the following conditions are met:

- Vehicle enters the intersection after the signal light had been red for a minimum elapsed time of 0.2 second, and
- The measured speed of violating vehicle is more than 15 mph (on straight-thru lane) or 13 mph (on left-turn lane).

Several pictures are taken to capture the vehicle crossing the stop bar after the red signal, license plate number of the vehicle, and the person who is driving the violating vehicle. The following steps are carried sequentially to process RLPEC pictures of a potential red light violator:

- Field service technicians from Affiliated Computer Services (ACS) (the company that maintains the photo enforcement cameras and processes all the red light violation data for the City of Sacramento) pick up images and data from the cameras.
- Pictures taken by the cameras are developed, digitized, and saved in a secure database, along with other information gathered by the cameras (e.g., date and time of the red light violation, vehicle speed of the violator, and elapsed time of the red signal).
- California Department of Motor Vehicles furnishes information on the registered owner of the violating vehicle, using license plate number captured by the photo enforcement cameras.
- Staff members at the Sacramento Police Department review pictures taken by the cameras and determine whether to issue violation citations. A violation citation will be issued only if evidence presented by the pictures indicates a definite red light violation.
- If a red light violation citation is issued, a copy of the citation is mailed to the violator. In addition, violation photos and other related data are uploaded onto a secure (i.e., password-protected) Web site to be viewed by the violator. After information viewing is completed, the City of Sacramento collects a red light violation fine from the violator.

Procedures described above reveal that not all pictures taken by the RLPECs lead to violation citations. No violation citation is issued if, for instance, pictures taken by the photo enforcement cameras do not clearly show that the vehicle crossed the stop bar after the onset of red signal, or the registered owner of the vehicle does not match the driver who ran the red light. Based on one estimate, only 35 percent of the pictures taken by Sacramento's cameras result in violation citations. Red light violation records used for this analysis only include the actual violation citations issued by the Sacramento Police Department. Consequently, red light violation rates estimated from Sacramento's data are likely to be lower than most violation rates reported by other studies (see Section 2).

3.3. Descriptive Statistics

This subsection presents all background statistics related to the Sacramento red light violation data used in this study. Pertinent information regarding the 11 signalized intersections equipped with RLPECs is also provided below.

3.3.1. General Information on Red Light Violation Records

With approval from the Sacramento Police Department, ACS prepared and provided to the Volpe Center data on more than 4 years (May 1999 to June 2003) of red light violation records collected from 11 RLPEC-equipped intersections. Figure 1 displays the relative locations of these 11 intersections in the City of Sacramento. Data from a total of 46,997 red light violation records were provided to the Volpe Center. Table 5 provides a distribution of these records and their collection periods (first and last red light violation record dates) by intersection location since not all 11 signalized intersections have a complete 4-year data set.

The following variables are included in Sacramento's red light violation data file:

- Intersection (and intersection code) where the violation occurred
- Date when the violation occurred
- Time when the violation occurred
- Age of the violator
- Gender of the violator
- Car (i.e., vehicle make) driven by the violator
- Model year of the vehicle driven by the violator
- Vehicle speed (i.e., measured speed) at the time of the violation
- Elapsed time from the onset of red signal until the time of the violation

All records in this data set have information on location of the violation, date of the violation, and time of the violation. However, information related to age and gender of the violator, type of vehicle driven by the violator, and year of the vehicle is missing from some violation records. Few records have no information on vehicle speed at the time of the violation.

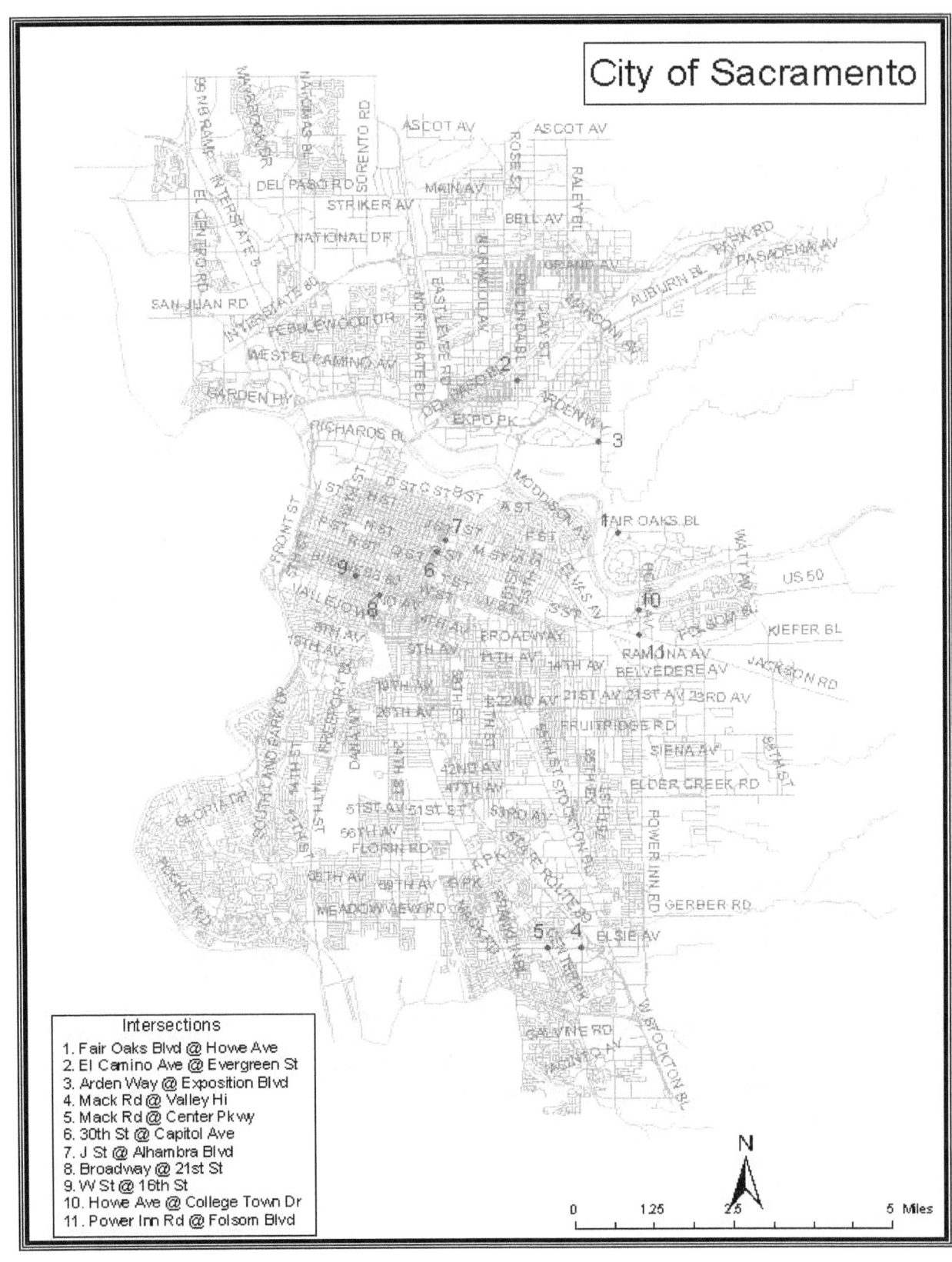

Figure 1. Location of Sacramento's 11 RLPEC-Equipped Intersections

Table 5. Distribution of Red Light Violation Records and their Collection Period

Signalized Intersection	No. of Red Light Violation Records	Date of the First Violation Record	Date of the Last Violation Record
Fair Oaks Boulevard and Howe Avenue	11,134	6/10/1999	6/21/2003
El Camino Avenue and Evergreen Street	6,167	6/4/1999	6/21/2003
Arden Way and Exposition Boulevard	1,820	4/25/2000	9/8/2002
Mack Road and La Mancha Way/Valley Hi Drive	2,408	5/26/1999	6/20/2003
Mack Road and Center Parkway	1,882	12/17/1999	6/20/2003
30th Street and Capitol Avenue	762	2/26/2000	7/12/2002
J Street and Alhambra Boulevard	5,475	2/29/2000	6/23/2003
Broadway and 21st Street	6,118	3/13/2000	12/4/2002
W Street and 16th Street [Highway 50 Exit at 16th Street]	5,592	6/30/2000	6/20/2003
Howe Avenue and College Town Drive	5,349	7/13/2000	8/9/2002
Power Inn Road and Folsom Boulevard	290	2/28/2002	3/28/2002
SUM	**46,997**	***	***

3.3.2. Information on RLPEC-Equipped Signalized Intersections

The Traffic Engineering Services at Sacramento Department of Public Works provided the Volpe Center with information related to traffic and infrastructure characteristics of the 11 RLPEC-equipped intersections. Table 6 quantifies these intersection characteristics.

Various resources were used to gather the traffic and infrastructure information presented in Table 6:

- Data on "Posted Speed Limit," "Yellow Time Duration," and "All Red Phase" were provided by ACS per direction from Sacramento's Traffic Engineering Services.
- Traffic volumes for the 11 intersections, based on the 24 hour average daily traffic (ADT) counts, were obtained from Sacramento Department of Public Works' website. Traffic counts for these intersections were carried out at various times, ranging from within a year to several years ago. 3 of the 11 intersections do not have complete count data for all approaches. Sacramento's traffic counts database can be accessed at www.pwsacramento.com/traffic/trafficcounts/index.cfm.
- Using the Geographic Information System (GIS) maps provided by Sacramento's Traffic Engineering Services, Volpe Center staff extracted the following information regarding these 11 intersections: "Average Intersection Gap," "Total Number of Lanes," and

"Average Lane Width." An example of the GIS map, for the intersection of Fair Oaks Boulevard and Howe Avenue, is shown in Appendix C with measurements of the intersection gap and lane width.

3.3.3. Frequency Distributions of Selected Variables

Frequency counts of the following variables are presented in this subsection:

- Age of the violator
- Gender of the violator
- Time (in hours) when the violation occurred
- Model Year of the vehicle driven by the violator
- Measured vehicle speed at the time of the violation
- Elapsed time from the onset of red signal until the time of the violation

In addition, the distribution of repeat red light offenders (i.e., drivers with more than 1 violation record) is also presented.

Age of the Violator: Distribution of red light violators, categorized by 7 age groups, is presented in Table 7. In addition, Table 7 includes data on the number of licensed drivers (LDs) in California (Federal Highway Administration, 2002), the total million vehicle miles traveled (MVMT) (Cerrelli, 1998), and relative ratios of red light violation (RLV) percentages by licensed driver percentages and total MVMT percentages. Relative ratios for $\frac{\% \text{ of RLV}}{\% \text{ of LD}}$ and $\frac{\% \text{ of RLV}}{\% \text{ of MVMT}}$ were plotted and presented in Figure 2. Both plotted lines in Figure 2 suggest that younger drivers under 30 years of age are more likely to run red lights compared to drivers in other age groups. The effect of age on red light violation behavior will be explored further in Section 4 of this report.

Gender of the Violator: Table 8 presents the distribution of red light violators by gender. The number of male and female licensed drivers in California (Federal Highway Administration, 2002), the total MVMT by male and female drivers (Cerrelli, 1998), and relative ratios of red light violation percentages by licensed driver percentages and total MVMT percentages are also listed in Table 8. Relative ratios of $\frac{\% \text{ of RLV}}{\% \text{ of LD}}$ show that male drivers might have a slightly higher tendency to run red lights. In contrast, relative ratios of $\frac{\% \text{ of RLV}}{\% \text{ of MVMT}}$ indicate that female drivers might be more likely to run red lights. The influence of gender on red light violation behavior will be further examined in Section 4 of this report.

Table 6. Traffic and Infrastructure Characteristics of the 11 RLPEC-Equipped Intersections

Signalized Intersection	Posted Speed Limit (mph)	Yellow Time Duration [1] (sec)	All Red Phase (sec)	Traffic Volume (ADT) [2]	Average Intersection Gap (ft)	Total Number of Lanes		Average Lane Width (ft)	
						North-South	East-West	North-South	East-West
Fair Oaks Boulevard and Howe Avenue	40	4.7 / 3.6	1.0 / 0.5	85,636	149	10	10	10	10
El Camino Avenue and Evergreen Street	35	3.9	0.0	29,563	84	3	6	17	12
Arden Way and Exposition Boulevard	40	4.7	0.5	83,765	118	8	6	12	13
Mack Road and La Mancha Way/Valley Hi Drive	35 to 40	3.6	0.0	43,765 [3]	142	9	9	10	10
Mack Road and Center Parkway	40	4.7	1.5	45,483	130	6	6	12	12
30th Street and Capitol Avenue	30	3.6	0.0	26,901	90	3	5	14	13
J Street and Alhambra Boulevard	25	3.6	0.0	38,793	83	4	6	8	13
Broadway and 21st Street	25	3.6	0.0	28,355	87	5	5	15	11
W Street and 16th Street [Highway 50 Exit at 16th Street]	35 [30 to 65]	3.9	0.0	46,473 [3]	92	5	3	9	15 [10]
Howe Avenue and College Town Drive	45	5.0	2.0	57,496 [3]	111	8	7	11	12
Power Inn Road and Folsom Boulevard	45	3.6	1.0	60,087	126	8	7	10	11

[1] Based on average value of recent yellow time durations that vary between peak and non-peak traffic hours.
[2] ADT = Average Daily Traffic, 24-Hour Counts.
[3] Estimated value based on ADT data from surrounding intersections.

Table 7. Distribution of Red Light Violation Records by Age Group

Age Group	No. of RLVs	% of RLVs	No. of LDs [1]	% of LDs	% of RLVs/ % of LDs	Total MVMT [2]	% of MVMT	% of RLVs/ % of MVMT
< or = to 19	1,668	4.27%	883,858	4.09%	**1.05**	83,169	3.96%	**1.08**
20 to 29	9,769	25.02%	3,925,985	18.16%	**1.38**	412,282	19.65%	**1.27**
30 to 39	9,448	24.20%	4,997,068	23.11%	**1.05**	539,014	25.68%	**0.94**
40 to 49	8,390	21.49%	4,797,117	22.18%	**0.97**	503,354	23.99%	**0.90**
50 to 59	5,381	13.78%	3,401,805	15.73%	**0.88**	288,915	13.77%	**1.00**
60 to 69	2,410	6.17%	1,883,240	8.71%	**0.71**	170,488	8.12%	**0.76**
> or = 70	1,979	5.07%	1,734,720	8.02%	**0.63**	101,386	4.83%	**1.05**
Sub-Total	**39,045**	***	***	***	***	***	***	***
Missing Data	*7,952*	***	***	***	***	***	***	***
Total	46,997	100.00%	21,623,793	100.00%	***	2,098,608	100.00%	***

[1] Number of licensed drivers in California, 2001.
[2] Total vehicle miles of travel in the U.S., in millions, 1996.

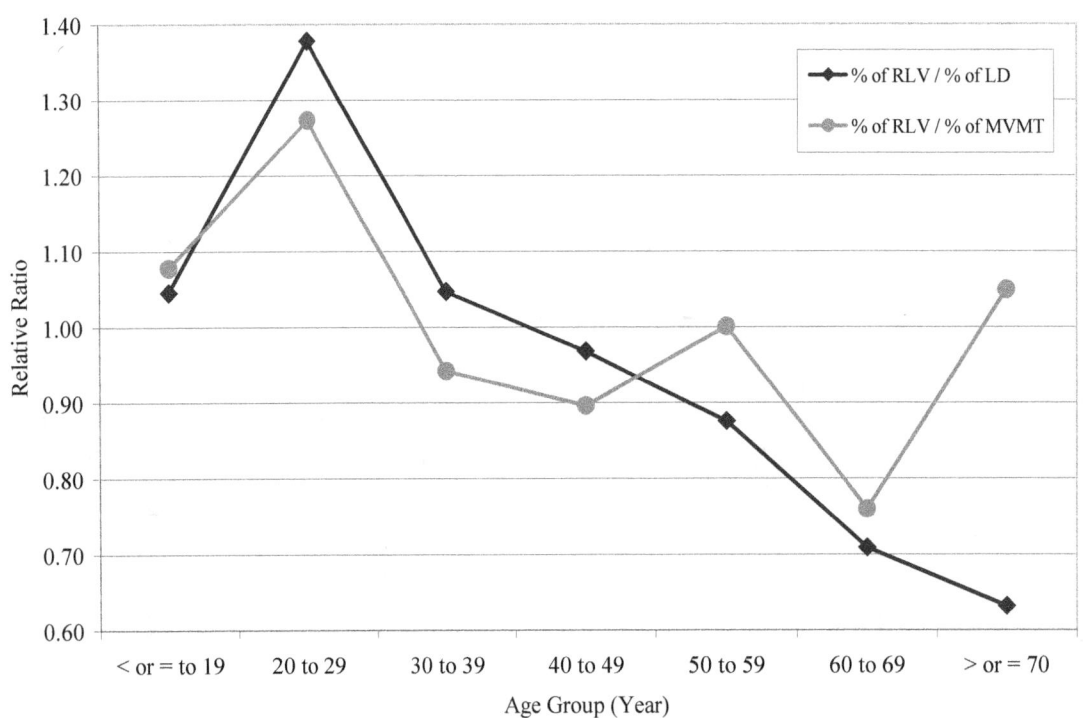

Figure 2. Normalized Red Light Violation Values by Age Group

Table 8. Distribution of Red Light Violation Records by Gender

Gender	No. of RLVs	% of RLVs	No. of LDs [1]	% of LDs	% of RLVs/ % of LDs	Total MVMT [2]	% of MVMT	% of RLVs/ % of MVMT
Male	24,798	55.51%	11,208,831	51.84%	**1.07**	1,317,941	62.80%	**0.88**
Female	19,876	44.49%	10,414,962	48.16%	**0.92**	780,667	37.20%	**1.20**
Sub-Total	**44,674**	***	***	***	***	***	***	***
Missing Data	*2,323*	***	***	***	***	***	***	***
Total	46,997	100.00%	21,623,793	100.00%	***	2,098,608	100.00%	***

[1] Number of licensed drivers in California, 2001.
[2] Total vehicle miles of travel in the U.S., in millions, 1996.

Violation Time: Figure 3 illustrates the distribution of Sacramento's red light violations by violation time (in hours). The overall trend shown in this figure is consistent with the expectation – most of red light violations occurred during the daytime hours when most urban driving is done (i.e., 7 a.m. to 7 p.m.). However, the highest count of red light violations during the time period from 2:00 p.m. to 2:59 p.m. is somewhat surprising. The relationship between time of day and red light violations will be studied further in the next section.

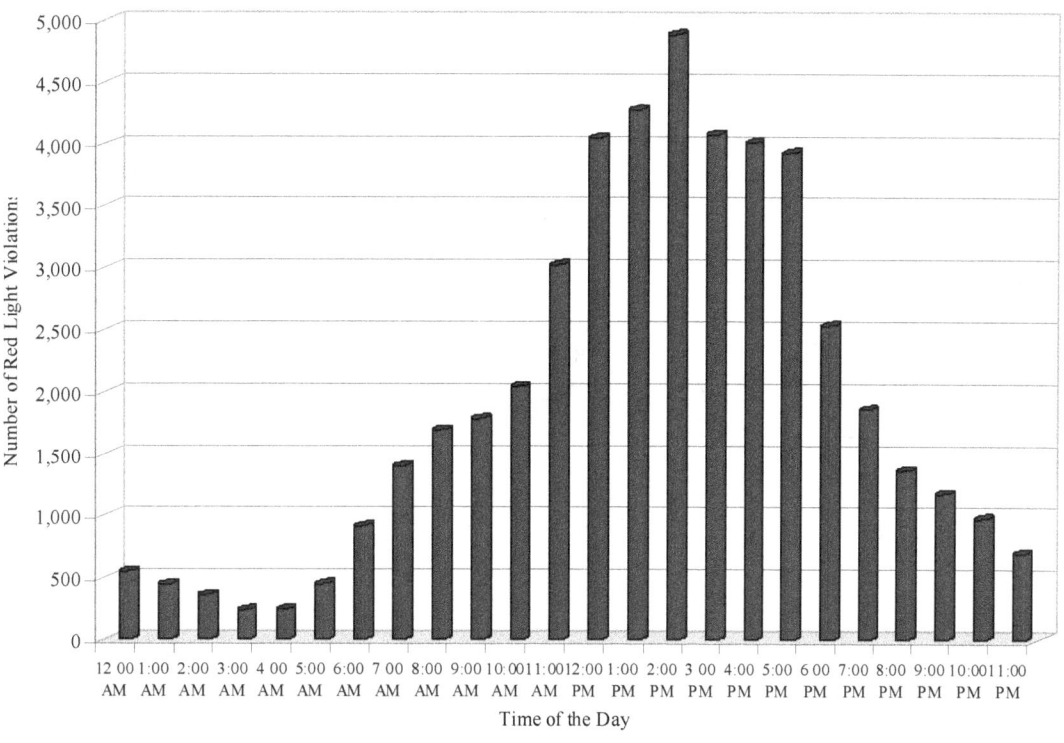

Figure 3. Distribution of Red Light Violations by Time of Day

Model Year of Violating Vehicle: About 99.1 percent of the 46,997 red light violation records from the City of Sacramento had information on model year of the violating vehicle. Figure 4 presents the distribution of red light violations per vehicle year.

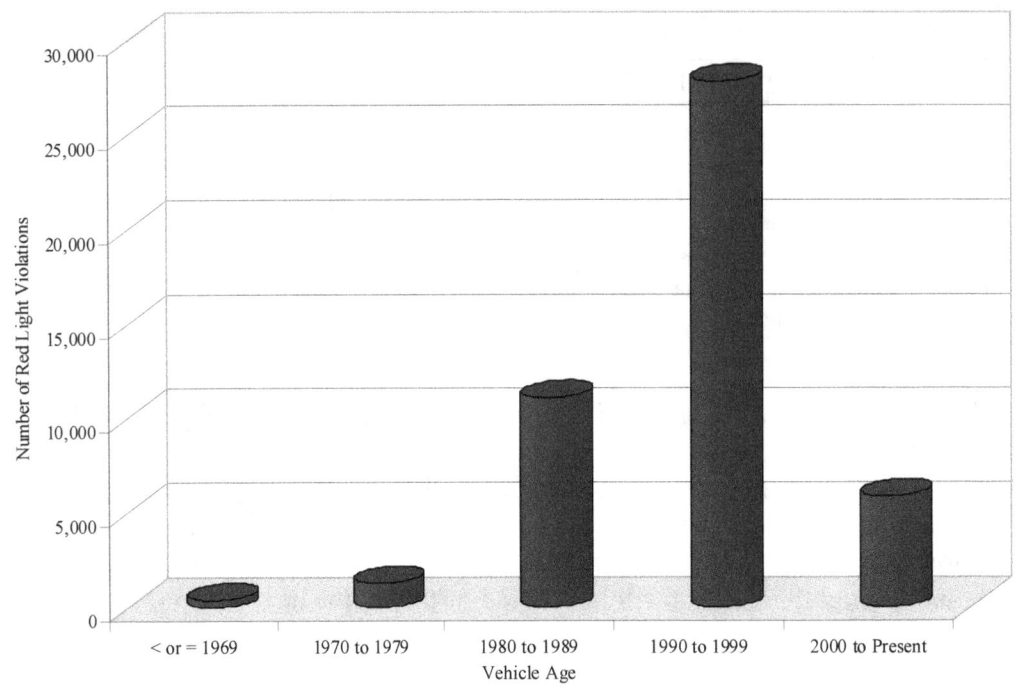

Figure 4. Distribution of Red Light Violation Records by Vehicle Year

Vehicle Speed at Time of Violation: A frequency distribution of the measured vehicle speeds at the time of red light violation is shown in Figure 5. A total of 46,993 out of the 46,997 records had data on measured vehicle speed at violation, ranging from 11 mph (2 records) to 87 mph (1 record). The three most frequent vehicle speeds at the time of violation were: 18 mph (1,762 or 3.7% of records), 17 mph (1,715 or 3.6% of records), and 19 mph (1,711 or 3.6% of records). The average red light violation speed was 31.6 mph. More than half (51.1%) of the drivers ran the red light at speeds of 30 mph or less. The highest posted speed limit among the 11 RLPEC-equipped intersections in Sacramento is 45 mph, and only 13.7 percent of the violators ran the red light at speeds higher than that limit. Figure 6 shows the percentage cumulative distribution of violations records by vehicle speed.

To examine speeding by the violating vehicles, data from violation records were grouped by individual intersection and posted speed limit (PSL). Speeds of violating vehicles were subtracted from corresponding PSLs at each intersection. Table 9 presents the distribution of red light violation records by "violation speed ≤ PSL" (i.e., not speeding) and "violation speed > PSL" (i.e., speeding) for the 11 RLPEC-equipped intersections. In addition, Figure 7 and Figure 8 illustrate respectively the percentage distribution and percentage cumulative distribution of red light violation records by vehicle speed minus the PSL (i.e., speeding measure). Appendix D presents these figures for individual intersections. Overall, about 56 percent of the violating vehicles were not speeding through the intersections at the time of the violation.

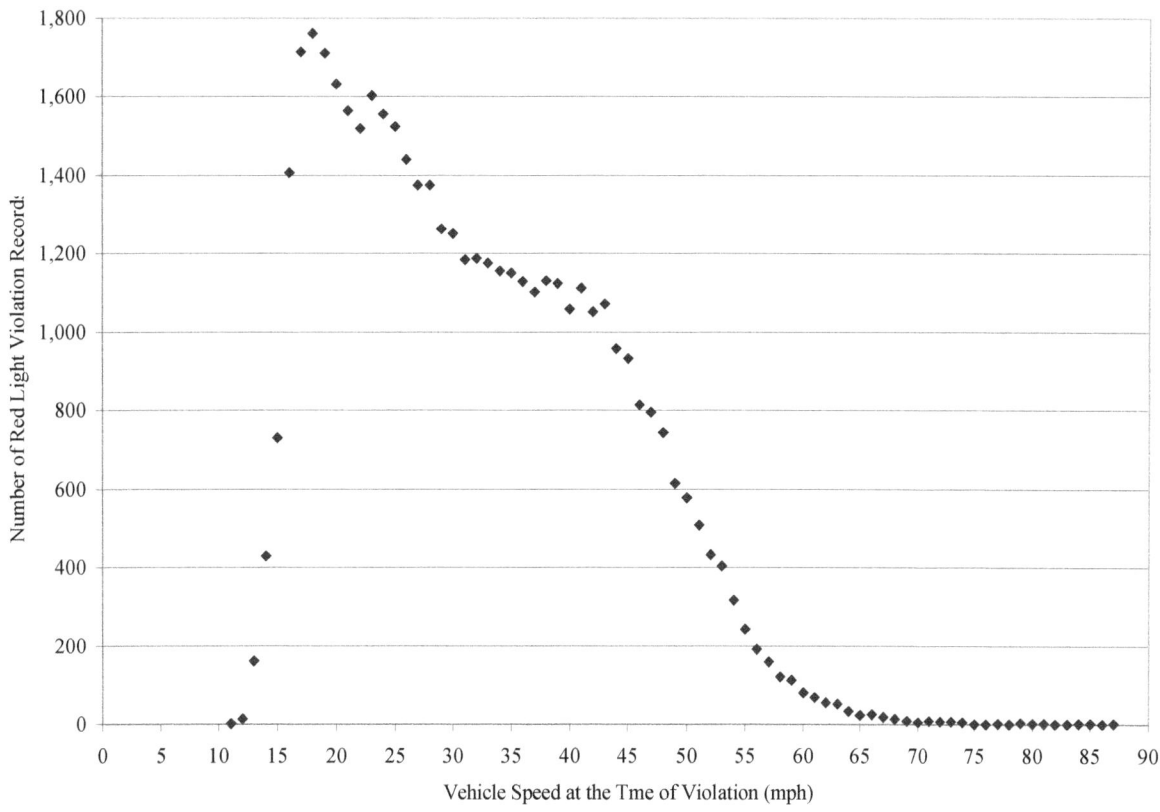

Figure 5. Distribution of Red Light Violation Records by Vehicle Speed

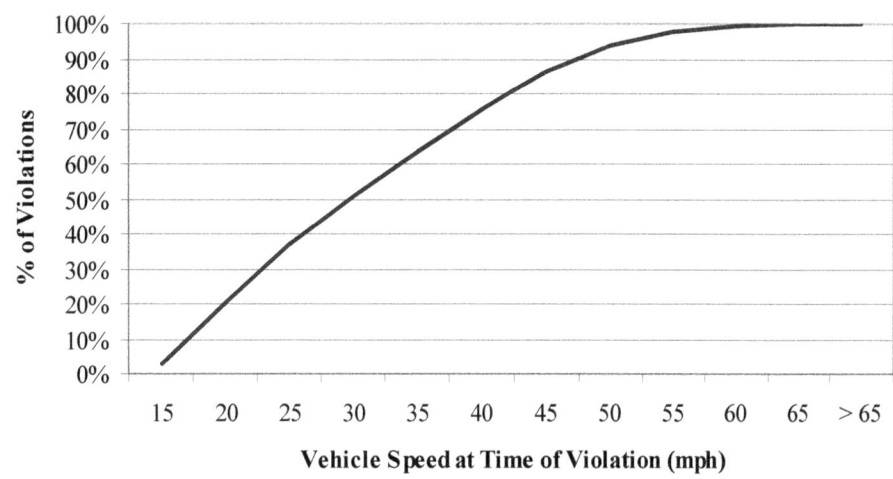

Figure 6. Percentage Cumulative Distribution of Red Light Violation Records by Vehicle Speed

Table 9. Distribution of Violation Records by Vehicle Speed in Comparison to the Posted Speed Limit

Signalized Intersection	PSL (mph)	Violation Speed ≤ PSL		Violation Speed > PSL		Total Red Light Violations
		Count	% of Total	Count	% of Total	
Fair Oaks Boulevard and Howe Avenue	40	10,535	94.6%	598	5.4%	11,133
El Camino Avenue and Evergreen Street	35	1,622	26.3%	4,545	73.7%	6,167
Arden Way and Exposition Boulevard	40	1,262	69.3%	558	30.7%	1,820
Mack Road and La Mancha Way/Valley Hi Drive	35	1,367	56.8%	1,039	43.2%	2,406
Mack Road and Center Parkway	40	526	27.9%	1,356	72.1%	1,882
30th Street and Capitol Avenue	30	396	52.0%	366	48.0%	762
J Street and Alhambra Boulevard	25	2,263	41.3%	3,212	58.7%	5,475
Broadway and 21st Street	25	1,193	19.5%	4,925	80.5%	6,118
W Street and 16th Street [Highway 50 Exit at 16th Street]	35	4,460	79.8%	1,131	20.2%	5,591
Howe Avenue and College Town Drive	45	2,458	46.0%	2,891	54.0%	5,349
Power Inn Road and Folsom Boulevard	45	290	100.0%	0	0.0%	290
Total	***	**26,372**	**56.1%**	**20,621**	**43.9%**	**46,993**

Examining the information presented in Table 9 and Figure 26 through Figure 36 (see Appendix D), the following is noted:

- Percentage distribution of "violation speed ≤ PSL" versus "violation speed > PSL" for the intersection of Mack Road and La Mancha Way/Valley Hi Drive is similar to the "overall" distribution for all 11 intersections.
- With limited amount of data, vehicle speeds for all red light runners at the intersection of Power Inn Road and Folsom Boulevard are less than or equal to the PSL.
- Intersections of Fair Oaks Boulevard and Howe Avenue, Arden Way and Exposition Boulevard, and W Street and 16th Street all have relatively low percentages of vehicles that ran the red light at speeds higher than the PSLs compared to the remaining intersections. According to Table 6, traffic volumes at these three intersections are all fairly high.

Additional analysis results regarding vehicle speed at the time of violation will be presented in Section 4.

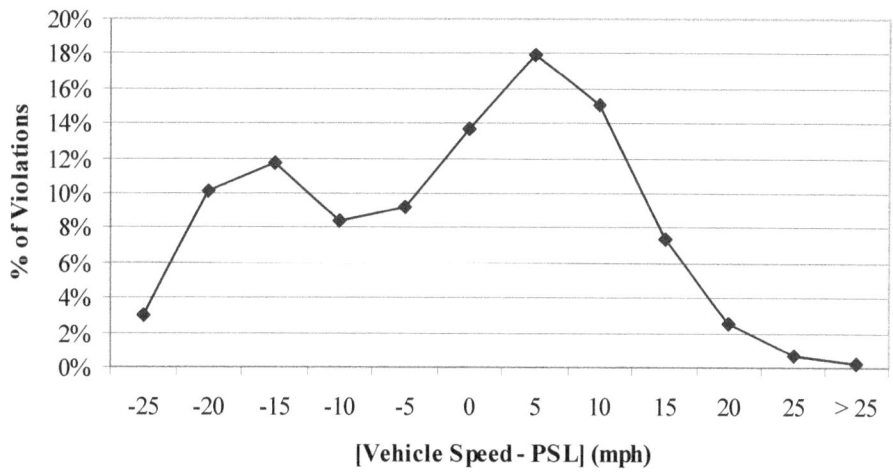

Figure 7. Percentage Distribution of Violation Records by [Vehicle Speed – PSL]

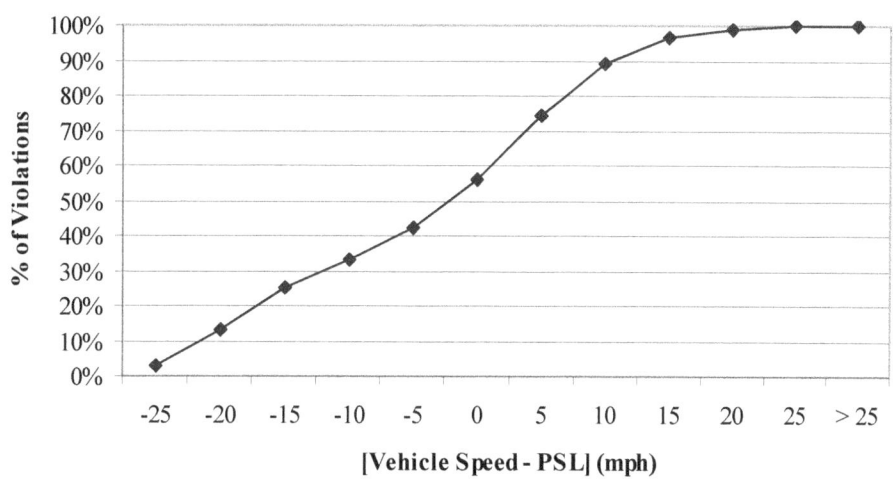

Figure 8. Percentage Cumulative Distribution of Records by [Vehicle Speed – PSL]

Elapsed Time from Onset of Red Signal until Time of Violation: Figure 9 presents a frequency distribution plot (on the logarithmic scale) of "elapsed time since red light onset" for all violation records. Information on red light elapsed time is available for every record in this Sacramento data set. This elapsed time is recorded to the nearest 0.1 second. RLPECs have captured drivers crossing intersections after the red light was initiated, from 0.2 second (minimum elapsed time to activate the enforcement camera, 6,381 records) to more than 30 seconds (434 records). Records with elapsed time greater than 30 seconds were not plotted in Figure 9. Figure 10 and Figure 11 illustrate respectively the percentage distribution and percentage cumulative distribution of red light violation records by elapsed time since red light onset.

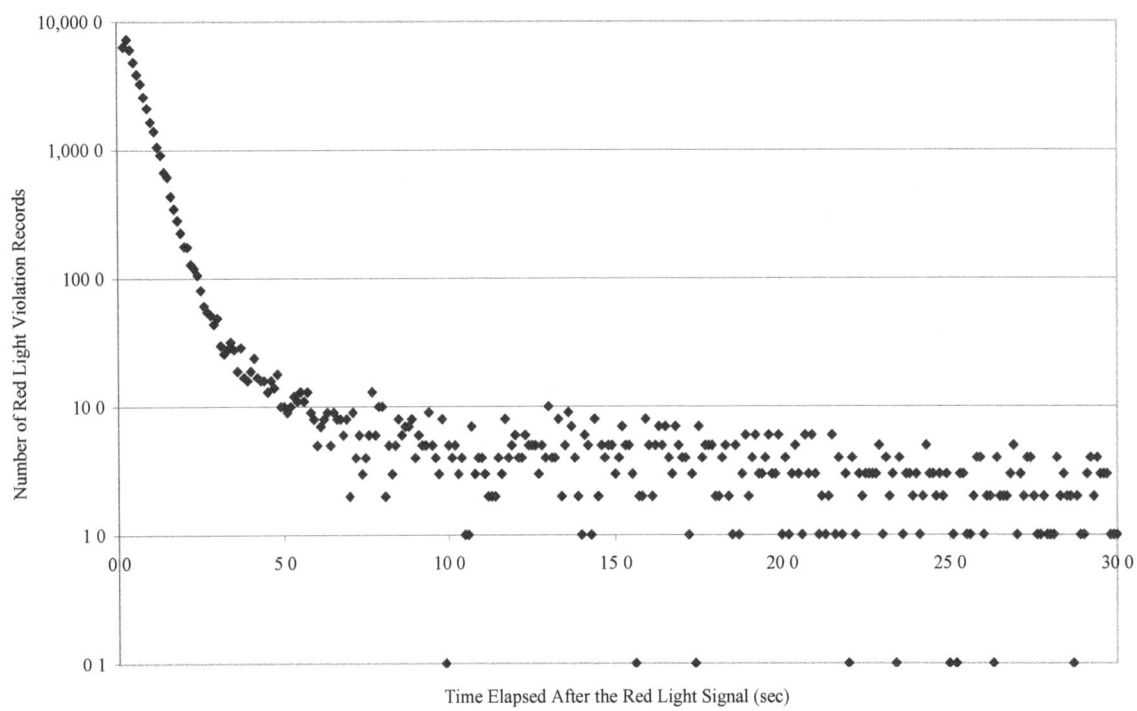

Figure 9. Distribution of Violation Records by Elapsed Time Since Red Light Onset (Vertical Axis in Logarithmic Scale)

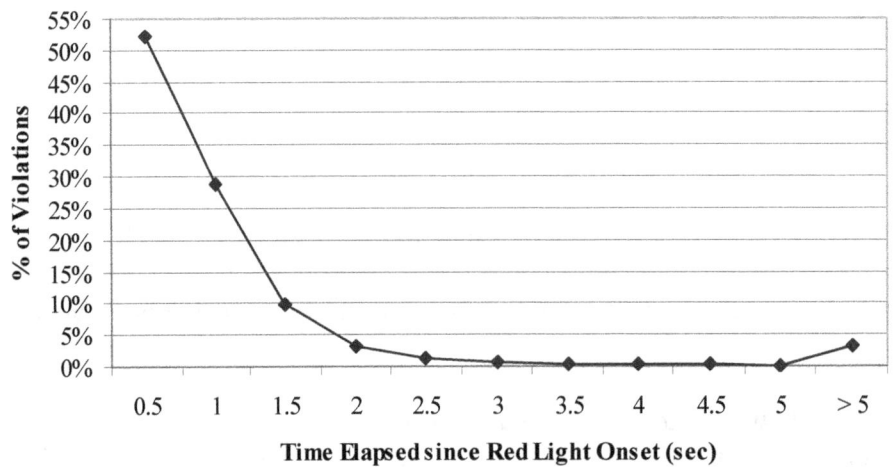

Figure 10. Percentage Distribution of Violation Records by Time Elapsed Since Red Light Onset

Lum and Wong (2003) separated red light violators into three groups according the "after-red time" (i.e., elapsed time when a driver goes through the intersection after the onset of red light): up to 2 seconds of after-red times, 2 to 5 seconds of after-red times, and more than 5 seconds of after-red times. Using the same categories, Table 10 lists the corresponding number of violation records from Sacramento's data set.

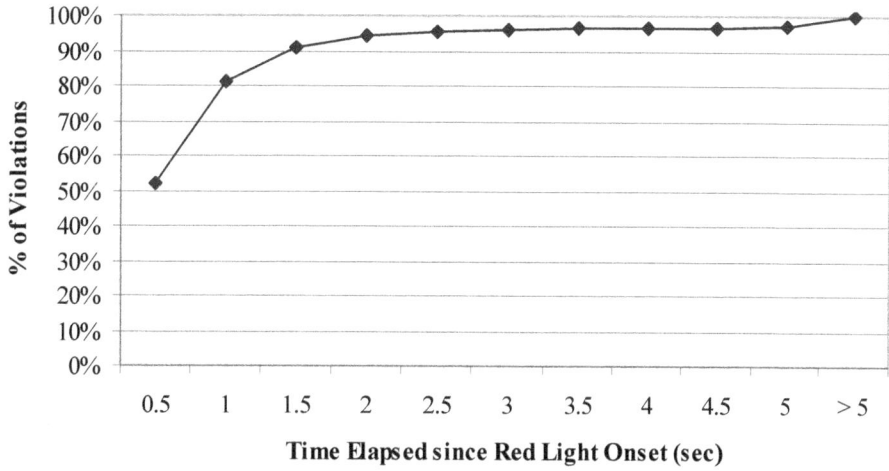

Figure 11. Percentage Cumulative Distribution of Violations by Time Elapsed Since Red Light Onset

Table 10. Distribution of Red Light Violation Records by Selected Categories of Time Elapsed Since Red Light Onset

After-Red Time	No. of Violation Records	Percent of Distribution	Cumulative Violation Records	Cumulative Percent
≤ 2.0 seconds	44,294	94.2%	44,294	94.2%
2.1 to 5.0 seconds	1,270	2.7%	45,564	97.0%
> 5.0 seconds	1,433	3.0%	46,997	100.0%

Information presented in Table 10 shows that more than 94 percent of red light violations in the City of Sacramento occurred within 2 seconds after the onset of the red light. The influence of the "dilemma zone" is probably one of the major reasons for such a high percentage of red light violators in this first group.

Vehicle Speed versus Time Elapsed since Red Light Onset: Table 11 provides values for the percentage cumulative distribution of violation records by time elapsed since red light onset for various vehicle speeds at time of violation. Over 95 percent of the violators who were traveling at speeds between 35 and 65 mph ran the red light within 2 seconds of red light onset. However, only 85 percent of the violators at speeds over 65 mph ran the red light within 2 seconds. Moreover, about 12 percent of the violators at speeds over 65 mph ran the red light after 5 seconds of its onset. About 95 percent of the violators traveling at speeds below 35 mph ran the red light within 2.5 seconds of its onset.

Vehicle Speed Minus Speed Limit versus Time Elapsed since Red Light Onset: Table 12 provides values for the percentage cumulative distribution of violation records by time elapsed since red light onset for various delta speeds (vehicle speed minus PSL) at time of violation. Over 95 percent of the violators traveling between 5 and 25 mph over the speed limit ran the red light within 2 seconds of its onset. Similar to the observation in Table 11, about 12 percent of the violators who were traveling higher than 25 mph over the speed limit ran the red light after 5 seconds of its onset. As seen in Table 11 and Table 12, there is no trend between "speeding" and time elapsed since red light onset except at very high speeds. The reader is cautioned, however, to the small sample of violation records at these very high speeds (i.e., vehicle speed > 65 mph and delta speed > 25 mph).

Table 11. Percentage Cumulative Distribution of Violations by Elapsed Time since Red Light Onset for Various Vehicle Speeds

Vehicle Speed (mph)	Time Elapsed since Red Light Onset (sec)											No. Records by Speed
	0.5	1	1.5	2	2.5	3	3.5	4	4.5	5	>5	
15	37.6%	67.7%	85.5%	93.6%	96.9%	97.8%	98.1%	98.1%	98.4%	98.5%	100.0%	1,334
20	44.8%	75.2%	87.8%	92.5%	94.7%	95.7%	96.1%	96.3%	96.5%	96.7%	100.0%	8,227
25	51.4%	81.4%	91.3%	94.2%	95.4%	95.9%	96.3%	96.6%	96.8%	96.9%	100.0%	7,763
30	55.7%	83.6%	92.0%	94.1%	94.9%	95.4%	95.7%	95.9%	96.1%	96.2%	100.0%	6,702
35	56.9%	84.1%	92.4%	94.8%	95.7%	96.2%	96.5%	96.8%	96.9%	97.1%	100.0%	5,851
40	56.0%	83.7%	92.3%	95.0%	96.0%	96.4%	96.6%	96.8%	97.0%	97.3%	100.0%	5,542
45	54.1%	82.9%	92.0%	95.0%	96.2%	96.7%	96.8%	97.1%	97.3%	97.3%	100.0%	5,123
50	54.1%	83.3%	92.6%	95.5%	96.5%	96.8%	97.0%	97.2%	97.4%	97.4%	100.0%	3,546
55	53.1%	82.4%	92.7%	95.2%	96.3%	96.5%	96.7%	96.8%	96.8%	97.0%	100.0%	1,902
60	54.6%	82.2%	92.5%	93.7%	94.7%	95.0%	95.5%	95.8%	95.9%	96.1%	100.0%	663
65	48.7%	83.6%	93.5%	95.7%	97.0%	97.8%	97.8%	98.3%	98.3%	98.3%	100.0%	232
>65	36.1%	68.5%	78.7%	85.2%	88.0%	88.0%	88.0%	88.0%	88.0%	88.0%	100.0%	108

Table 12. Percentage Cumulative Distribution of Violations by Elapsed Time since Red Light Onset for Various Delta Speeds

Delta Speed (mph)	Time Elapsed since Red Light Onset (sec)											No. Records by Delta Speed
	0.5	1	1.5	2	2.5	3	3.5	4	4.5	5	>5	
-25	38.8%	69.4%	86.5%	94.4%	97.6%	98.5%	98.8%	98.8%	99.0%	99.1%	100.0%	1,430
-20	42.3%	73.3%	87.9%	93.8%	96.5%	97.4%	97.9%	98.1%	98.2%	98.3%	100.0%	4,774
-15	47.3%	77.2%	87.4%	91.2%	92.9%	93.9%	94.5%	94.8%	95.0%	95.2%	100.0%	5,528
-10	49.9%	79.3%	89.8%	92.7%	93.7%	94.3%	94.8%	95.1%	95.4%	95.6%	100.0%	3,933
-5	50.6%	78.1%	87.7%	90.8%	92.3%	93.2%	93.6%	93.8%	94.2%	94.8%	100.0%	4,306
0	52.8%	81.6%	90.7%	93.4%	94.7%	95.3%	95.6%	96.0%	96.3%	96.5%	100.0%	6,401
5	56.9%	84.5%	93.1%	95.5%	96.3%	96.7%	96.8%	97.0%	97.2%	97.2%	100.0%	8,419
10	58.5%	87.0%	95.1%	97.0%	97.6%	97.9%	97.9%	98.0%	98.0%	98.1%	100.0%	7,102
15	59.0%	87.4%	95.5%	97.6%	98.1%	98.3%	98.4%	98.4%	98.5%	98.5%	100.0%	3,480
20	54.7%	85.1%	95.5%	97.9%	98.7%	98.8%	99.1%	99.1%	99.1%	99.1%	100.0%	1,191
25	51.5%	82.4%	93.8%	96.7%	98.0%	98.0%	98.0%	98.0%	98.0%	98.0%	100.0%	307
>25	32.5%	66.7%	77.5%	83.3%	87.5%	87.5%	87.5%	88.3%	88.3%	88.3%	100.0%	120

Delta Speed= Vehicle speed – speed limit

Repeat Red Light Offenders: Table 13 shows the distribution of repeat red light offenders. ACS determined these counts by matching vehicle license plate numbers. To protect the identity of red light offenders, vehicle license plate numbers were not part of the data given to the Volpe Center. Consequently, no further analysis on repeat offenders could be performed except for the information presented in Table 13. Even though there are close to 2,000 repeat red light violators, such number is still a relatively small fraction or 4 percent of the entire Sacramento data set.

Table 13. Statistics on Repeat Red Light Offenders

No. of Repeated Offenses	No. of Repeat Offenders (via Vehicle Plate Matching)	Repeat Offenders vs. Total No. of Violators	No. of Red Light Violation Records	Repeated Violation Records vs. Total Violation Records
2	1,704	3.79%	3,408	7.25%
3	137	0.30%	411	0.87%
4	17	0.04%	68	0.14%
5	2	0.004%	10	0.02%
7	1	0.002%	7	0.01%
Total	1,861	4.14%	3,904	8.31%

Total Number of Red Light Violations from the Database =	46,997
Number of Violators with 1 Red Light Violation =	43,093
Total Number of Violators from the Database =	44,954
Percent of Violators with 1 Red Light Violation =	95.86%

3.4. Estimates of Red Light Violation and Crash Rates

Table 14 presents estimates of red light violation rates from the 11 RLPEC-equipped intersections in the City of Sacramento. These violation rates were calculated as a function of the following factors based on available information gathered for this study:

- *Number of red light violation records*: Red light violation records for each intersection were tabulated on a monthly basis. Some intersections have incomplete monthly violation records, possibly caused by malfunction of the red light camera. Monthly red light violation rates were calculated for each intersection. Values listed in Table 14 were the average rates from the sum of monthly rates, excluding those with incomplete monthly records.
  Traffic volumes at intersections: Monthly traffic volumes at each intersection were calculated using the 24 hour ADT counts, obtained from Sacramento Department of Public Works' website. The City of Sacramento conducted traffic counts at these intersections during various times; some count data were relatively recent while others were collected several years ago. It is assumed that ADTs at these intersections remained similar in the past several years. Three intersections do not have complete counts; hence,

approximate ADT values for these intersections were generated based on available traffic counts from surrounding intersections. Monthly traffic volumes, from May 1999 to June 2003, were calculated based on the number of weekdays and weekends in each month. Weekend ADTs (Saturday and Sunday) were assumed at 25 percent of the weekday ADTs for all intersections.

Table 14. Estimates of Red Light Violation Rates From 11 RLPEC-Equipped Intersections

Location Code	Signalized Intersection	No. of RLVs	ADT [1,2]	Records Used to Estimate Violation Rates		Violation Rate Estimates from Issued Citations [4] (Per 1,000 Vehicles)	Violation Rate Estimates from Photos Taken by the Enforcement Camera [5] (Per 1,000 Vehicles)
				Duration of the Records [3]	No. of Records		
361304	Fair Oaks Boulevard and Howe Avenue	11,134	85,636	04/2000 - 09/2002	9,667	**0.157**	*0.449*
362301	El Camino Avenue and Evergreen Street	6,167	29,563	06/1999 - 08/2002	5,384	**0.195**	*0.557*
363201	Arden Way and Exposition Boulevard	1,820	83,765	01/2000 - 12/2000	1,538	**0.064**	*0.183*
363302	Mack Road and La Mancha Way/Valley Hi Drive	2,408	43,765	06/1999 - 06/2000	2,031	**0.149**	*0.426*
364301	Mack Road and Center Parkway	1,882	45,483	01/2000 - 06/2000 & 01/2001 - 12/2001	1,609	**0.082**	*0.234*
365101	30th Street and Capitol Avenue	762	26,901	03/2000 - 06/2000 & 11/2001 - 02/2002	681	**0.133**	*0.380*
365301	J Street and Alhambra Boulevard	5,475	38,793	03/2000 - 08/2002	4,713	**0.169**	*0.483*
365401	Broadway and 21st Street	6,118	28,355	04/2000 - 08/2002	5,789	**0.294**	*0.840*
366402	W Street and 16th Street [Highway 50 Exit at 16th Street]	5,592	46,473	07/2000 - 08/2002	5,049	**0.174**	*0.497*
367102	Howe Avenue and College Town Drive	5,349	57,496	08/2000 - 06/2002	5,042	**0.159**	*0.454*
367601	Power Inn Road and Folsom Boulevard	290	60,087	03/2002	280	**0.198**	*0.566*

[1] ADT at each intersection is used to estimate the monthly traffic volume.
[2] ADT during the weekend (Saturday and Sunday) is assumed to be 25 percent of the ADT during weekdays (Monday to Friday) when calculating the monthly traffic volume.
[3] Violation rate calculations included only those months where red light violation records were available for the entire month.
[4] This red light violation rate is calculated based on the data set presented in this report.
[5] This red light violation rate is generated assuming only 35 percent of the photos taken by Sacramento's red light enforcement cameras turn into violation citations (see Subsection 3.2).

- *Violation rate estimates from issued citations versus violation rate estimates from photos taken by the red light cameras*: Two sets of red light violation rates were shown in Table 14. The first group of violation rates was calculated based on the red light running records used in this study. The second group of violation rates was approximated based on the assessment that only approximately 35 percent of the photos taken by Sacramento's cameras turn into violation citations.

The red light violation rates in Table 14, as calculated from issued violation citations, range from 0.064 violation per 1,000 crossing vehicles to 0.294 violation per 1,000 crossing vehicles. In comparison to red light violation rates reported by other studies (see Table 3), these estimates of red light violation rates are quite low, probably due to the following reasons:

- The 46,997 red light violation records used to estimate the violation rates were actual citations issued by Sacramento Police Department. As discussed in subsection 3.2, Sacramento Police Department carefully reviews all photos taken by the enforcement cameras and would only issue violation citations when certain criteria were met. As a result, approximately 35 percent of the photos taken by Sacramento's RLPECs result in violation citations. Consequently, estimates of red light violation rates from this data set are lower and more conservative.
- Sacramento's red light violation records analyzed in this study are an accumulation of several years of data instead of observations made in few weeks or months. These violation records were collected from an ongoing red light violation prevention program instead of a temporary/pilot study. Therefore, the Sacramento data set is comprehensive and stable compared to data gathered from a short-term study (brief data collection period and from few intersections with frequent red light running activities).
- Some assumptions made in this study might have been too conservative and thus resulted in low estimates of red light violation rates.

No obvious trend has been observed between the posted speed limit, yellow time, all-red time, traffic volume, intersection gap, number of lanes, and lane width versus the red light violation rates. Red light violation rates (calculated from issued violation citations) for these 11 intersections can be categorized as follows:

- Two of the intersections (Arden Way and Exposition Boulevard & Mack Road and Center Parkway) have violation rates that are lower than 0.1 violation per 1,000 crossing vehicles. These two intersections have the same PSL, similar clearance intervals, and similar number of lanes. One intersection has larger ADT but smaller average intersection gap.
- Eight of the intersections have red light running rates ranging from 0.1 violation per 1,000 crossing vehicles to 0.2 violation per 1,000 crossing vehicles. These intersections have various PSLs, clearance intervals, traffic volumes, average intersection gaps, and number of lanes.
- Only one intersection was estimated to have a violation rate greater than 0.2 violation per 1,000 crossing vehicles (Broadway and 21st Street). This intersection has a PSL of 25 mph with 3.6 seconds of yellow time and no all-red phase. This can be considered as one of the "smaller" intersections from the eleven examined in this report with low ADT, average intersection gap, and number of lanes.

Finally, Table 15 summarizes vehicle collision records at the 11 RLPEC-equipped intersections. These collision records were checked against the respective red light violation records and violation rates but no obvious relationship could be established. Excluding the intersection with zero collision due to small sample of violation records, crash rate estimates vary between 1 and 9 crashes per 1,000 red light violations. Overall, the crash rate is estimated at about 5 crashes per 1,000 red light violations based on this Sacramento data set.

Table 15. Collision Rates from Sacramento's RLPEC-Equipped Intersections

Location Code	Records Used to Estimate Violation Rates		Violation Rate Estimates from Issued Citations [2] (Per 1,000 Vehicles)	No. of Collision Records [3] [4]	Collision Normalized by Red Light Violation Records
	Duration of Records [1]	No. of Records			
361304	04/2000 - 09/2002	9,667	0.157	54	0.0056
362301	06/1999 - 08/2002	5,384	0.195	49	0.0091
363201	01/2000 - 12/2000	1,538	0.064	5	0.0033
363302	06/1999 - 06/2000	2,031	0.149	18	0.0089
364301	01/2000 - 06/2000 & 01/2001 - 12/2001	1,609	0.082	13	0.0081
365101	03/2000 - 06/2000 & 11/2001 - 02/2002	681	0.133	6	0.0088
365301	03/2000 - 08/2002	4,713	0.169	8	0.0017
365401	04/2000 - 08/2002	5,789	0.294	15	0.0026
366402	07/2000 - 08/2002	5,049	0.174	5	0.0010
367102	08/2000 - 06/2002	5,042	0.159	41	0.0081
367601	03/2002	280	0.198	0	0.0000
Total	***	41,783	***	214	***

[1] Violation rates included only those months where red light violation records were available for the entire month.
[2] This red light violation rate is calculated based on the data set presented in this report.
[3] Sacramento's Traffic Engineering Services provided information from the "Traffic Collision History Report".
[4] Only collisions that were in the "Duration of the Records" were counted.

4. STATISTICAL MODELING OF RED LIGHT VIOLATION DATA

This section presents results from statistical modeling of Sacramento's red light violation data. Two logistic regression models were developed to examine the influence of age, gender, violation time, vehicle year, and violation location on two *dependent* variables: (1) vehicle speed at the time of violation and (2) elapsed time between the onset of red signal and the time of violation.

4.1. Logistic Modeling Method

Logit models were developed to assess the effect of age, gender, violation time, vehicle year, and violation location on two *dependent* variables: (1) violators' vehicle speed when they ran the red light (i.e., ≤ PSL versus > PSL) and (2) elapsed time since red light onset (i.e., ≤ 2 seconds versus > 2 seconds). Since the input values for the two *dependent* variables (usually denoted as y_i) are dichotomous, logistic regression analysis is employed to understand the effect of various *independent* (or *explanatory*, usually denoted as x_i) variables on the dependent variables.

The general expression for the logit equation is:

$$p_i = \frac{e^{(\alpha + \beta_1 x_{i1} + \beta_2 x_{i2} + \ldots + \beta_k x_{ik})}}{1 + e^{(\alpha + \beta_1 x_{i1} + \beta_2 x_{i2} + \ldots + \beta_k x_{ik})}} \tag{1}$$

where

p_i = Probability of y_i equals to 1 (1 = violator's vehicle speed > PSL *or* elapsed time since red light onset > 2 seconds when violator ran the intersection in this study)
i = 1, ..., n individuals
α = Constant term
$\beta_1 \ldots \beta_k$ = Coefficients corresponding to explanatory variables $x_1 \ldots x_k$
$x_1 \ldots x_k$ = Explanatory variables

Equation (1) can be simplified further by dividing both the numerator and denominator by the numerator itself:

$$p_i = \frac{1}{1 + e^{-(\alpha + \beta_1 x_{i1} + \beta_2 x_{i2} + \ldots + \beta_k x_{ik})}} \tag{2}$$

One very useful piece of information that can be easily generated from the logistic regression analysis is *odds ratios*. The odds of an event is the ratio of the expected number of times that an event will occur to the expected number of times it will not occur. The relationship between probabilities and odds can be expressed as:

$$O_i = \frac{p_i}{1 - p_i} \tag{3}$$

where

O_i = Odds of an event

Logistic regression analysis results, including odds ratios, regarding the influence of various explanatory variables on violators' vehicle speed at time of violation and elapsed time since red light onset are presented respectively in subsections 4.2 and 4.3 below. Logistic regression analysis of the Sacramento red light violation data was performed using the SAS software.

4.2. Factors with Significant Influence on Violators' Vehicle Speed

This subsection presents the results of logistic regression analysis using "vehicle speed at time of violation" as the dependent variable. This dependent variable has two categories of alternatives (or binary responses): (1) vehicle speed less than or equal to PSL (not speeding) and (2) vehicle speed greater than PSL (speeding).

Table 16 lists the dependent and explanatory variables used in the logistic regression analysis of vehicle speed at time of red light violation. Few items are worthy to point out regarding the explanatory variables shown in Table 16:

- After several trials using various formats, the breakdown for the age variable was set as: younger drivers between 15 to 29 years old, middle-age drivers between 30 to 59 years old, and older drivers at 60 years old or more.
- Variable codes "Time 1" to "Time 4" are the 4 *dummy variables* representing the four time periods when drivers ran the red light. In the logistic analysis, "Time 4" (from 8 p.m. to 5 a.m.) was selected as the reference variable during the data regression process because this is the "off-peak" period for driving with low traffic volume. In addition, red light violations from 8 p.m. to 5 a.m. were considerably lower compared to other hours of the day, according to Figure 3.
- The following logic was used to divide the "VehYr" variable into three categories: cars made 10 or more years ago were considered as "old cars," cars made on or after year 2000 were considered as "new cars," and any cars made before year 2000 but less than 10 years were placed in the "medium-age car" category.
- Variable codes "LocCode 1" to "LocCode 11" are the *dummy variables* representing the eleven Sacramento intersections used in the data analysis. "LocCode 4" (Mack Road and La Mancha Way/Valley Hi Drive) is used as the reference variable during the logistic regression analysis. "LocCode 4" is selected as the reference because the distribution of "violations occurred at speeds ≤ PSL" versus "violations occurred at speeds > PSL" at this intersection is similar to the overall distribution for the entire dataset (see Table 9).

Using "Speed2" as the dependent variable and the explanatory variables listed in Table 16, numerous runs of logistic regression were carried out to find a most suitable logit model that portrays the relationship between "vehicle speed at time of violation" and significant explanatory variables. In addition, several statistical tests were conducted to ensure things such as multicollinearity (occurs when there are strong linear dependencies among the explanatory variables) do not affect the robustness of the final model.

Table 16. Dependent and Explanatory Variables Used in Logistic Regression Analysis

Variable Description	Variable Code	Variable Type	Data Format	SAS Code
Speed of the Vehicle at Time of Violation, MPH	*Speed2*	*Dependent*	*Violation Speed ≤ Posted Speed Limit*	*0*
			Violation Speed > Posted Speed Limit	*1*
Age of the Red Light Violator	Age	Explanatory	Younger, 15 to 29 Years Old	1
			Middle-Age, 30 to 59 Years Old	2
			Older, 60 Years or Older	3
Gender of the Red Light Violator	Gender	Explanatory	Male	0
			Female	1
Time of the Red Light Violation	Time 1	Explanatory	6 a.m. to 9 a.m.	1
	Time 2		10 a.m. to 3 p.m.	2
	Time 3		4 p.m. to 7 p.m.	3
	Time 4		8 p.m. to 5 a.m.	4
Age of the Vehicle Driven by Violator	VehYr	Explanatory	Old, Made in or Before 1993	1
			Medium, Made from 1994 to 1999	2
			New, Made in or After 2000	3
Location Code for the Signalized Intersection	LocCode 1	Explanatory	Fair Oaks Boulevard and Howe Avenue	1
	LocCode 2		El Camino Avenue and Evergreen Street	2
	LocCode 3		Arden Way and Exposition Boulevard	3
	LocCode 4		Mack Road and La Mancha Way/Valley Hi Drive	4
	LocCode 5		Mack Road and Center Parkway	5
	LocCode 6		30th Street and Capitol Avenue	6
	LocCode 7		J Street and Alhambra Boulevard	7
	LocCode 8		Broadway and 21st Street	8
	LocCode 9		W Street and 16th Street [Highway 50 Exit at 16th Street]	9
	LocCode 10		Howe Avenue and College Town Drive	10
	LocCode 11		Power Inn Road and Folsom Boulevard	11

After numerous iterations of logistic regression model runs, testing various combinations of explanatory variables (including some interaction terms such as "Age*Gender" and "Age*Time 1"), a final logit model was selected with "Age," "Time," and "LocCode" included in the model as the explanatory variables. Main effects "Gender" and "VehYr" did not show any significant influence on "Speed2" (p-values > 0.05); hence, they were not included in the final model. In addition, none of the interaction terms showed significant effects toward "Speed2"; consequently, they were not part of the final model.

Estimated coefficients, standard errors, p-values, and odds ratios for all the explanatory variables in the final logit model are presented in Table 17. Out of the possible 46,997 records, 39,042 records were used for this logistic regression analysis (7,955 records were not used due to missing values for either the dependent or the explanatory variables). The *generalized R^2* and *max-rescaled R^2* values for this model, both measuring the predictive power of the model, are 0.338 and 0.453 respectively. The R^2 values are not high but adequate, considering the fact that close to 40,000 data points were utilized in the logistic regression.

Table 17. Estimation Results for the Binary Logit Model with "Speed2" as Dependent Variable

Variable Code	Estimated Coefficient	Standard Error	p-value	Odds Ratio		1/Odds Ratio
Age	-0.3883	0.0217	< 0.0001		0.678	1.475
Time 1	-0.1986	0.0490	< 0.0001	*versus Time 4*	0.820	1.220
Time 2	-0.8645	0.0392	< 0.0001	*versus Time 4*	0.421	2.375
Time 3	-0.7142	0.0417	< 0.0001	*versus Time 4*	0.490	2.043
LocCode 1	-2.4797	0.0709	< 0.0001	*versus LocCode 4*	0.084	11.938
LocCode 2	1.4583	0.0605	< 0.0001	*versus LocCode 4*	4.299	0.233
LocCode 3	-0.4206	0.0775	< 0.0001	*versus LocCode 4*	0.657	1.522
LocCode 5	1.2514	0.0776	< 0.0001	*versus LocCode 4*	3.495	0.286
LocCode 6	0.3613	0.0973	0.0002	*versus LocCode 4*	1.435	0.697
LocCode 7	0.7995	0.0589	< 0.0001	*versus LocCode 4*	2.224	0.450
LocCode 8	1.8422	0.0620	< 0.0001	*versus LocCode 4*	6.311	0.158
LocCode 9	-0.9924	0.0623	< 0.0001	*versus LocCode 4*	0.371	2.698
LocCode 10	0.5550	0.0586	< 0.0001	*versus LocCode 4*	1.742	0.574
LocCode 11	-13.7451	80.5027	0.8644			

The estimated coefficient for variable "Age" is a negative value, signifying that as the age of the violator increases, the probability of running the red light while speeding decreases. The inverse value of the odds ratio for "Age," 1.475, implies that the predicted odds of a younger driver running a red light while speeding is about 1.5 times the odds of a middle-age driver running a red light while speeding. The same odds ratio also applies when the age category goes from "middle-age" to "older."

Analyzed results for the "Age" variable are consistent with our expectation. Research results from other studies have found that younger motorists drive more aggressively and are more likely to take driving risks compared to older drivers. Hence, we would expect that there are more younger drivers who would attempt to "beat the red light" by going through the intersection at high speeds that are greater than the posted speed limit.

The estimated coefficients for "Time 1," "Time 2," and "Time 3" are all negative values. The negative coefficients suggest that drivers who run the red light in "Time 1," "Time 2," and "Time 3" have a lower probability of speeding through intersections than those who commit red light violations in "Time 4." Hours represented by "Time 4" are from 8 p.m. to 5 a.m., typically considered as non-peak hours with low traffic volumes. Red light violations occurring in this time period are relatively low, especially from midnight to 5 a.m.. However, a combination of non-peak hour and low traffic volume may encourage violators to run the red light at speeds greater than the posted speed limit, especially the intentional violators.

According to the inverse values of the odds ratio presented in Table 17, red light violators in "Time 4" are 1.2, 2.4, and 2.0 times more likely to go through the intersection at speeds above the posted speed limit than those who ran a red light in "Time 1," "Time 2," and "Time 3," respectively. "Time 2," from 10 a.m. to 3 p.m., has the highest number of total red light violations of any time periods and "Time 1" and "Time 3" represent the a.m. and p.m. peak hours respectively; however, it appears that many drivers who ran red light in these three time periods have done so at speeds lower than the posted speed limit. Possible reasons for such phenomenon include:

- Violators might be "forced" to cross the intersection to avoid being rear-ended;
- Violators might try to race through the intersection but could not due to heavy surrounding traffic or don't want to get caught speeding;
- A combination of shorter signal timing cycle and high traffic volume might have created many dilemma zones that caused indecisive drivers to violate the red light.

The estimated coefficients for "LocCode 1" through "LocCode 11" are a mix of positive and negative values. "LocCode 11" is considered as an insignificant variable because its p-value is greater than 0.05. However, the estimated value for "LocCode 11" was still presented in Table 17 for informational purpose.

Compared to the reference variable, "LocCode 4," motorists who ran red light at "LocCode 1," "LocCode 3," and "LocCode 9" have a lower probability of speeding through these intersections. Positive coefficients for "LocCode 2," "LocCode 5," "LocCode 6," "LocCode 7," "LocCode 8," and "LocCode 10" suggest that red light violators at these locations are more likely to be speeding through intersections compared to "LocCode 4."

Examining the characteristics of these intersections, a common factor for "LocCode 1," "LocCode 3," and "LocCode 9" is identified: high traffic volumes pass through these intersections on a daily basis. "LocCode 1" and "LocCode 3" are two key junctions in Sacramento and one approach of "LocCode 9" is the off-ramp of a highway; hence, many vehicles pass through these intersections daily. Motorists who ran the red light at these three intersections might have had difficulty "speeding through" due to heavy surrounding traffic;

consequently, violators at these intersections are less likely to run red light at speeds higher than the posted speed limit compared to the reference intersection, "LocCode 4."

A logistic regression model presented in this subsection illustrated the influence of "Age," "Time," and "LocCode" on the speed of the violating vehicle. The results of this model allow us to better understand certain aspects of driver behavior around intersections, which will be useful to the design and development of effective vehicle-intersection cooperative signal violation warning systems.

4.3. Factors with Significant Influence on Time Elapsed since Red Light Onset

Lum and Wong (2003) suggested that most violators who ran through intersections after the onset of red light for more than 2 seconds were *deliberately* running the red light. Information presented in Table 10 and Figure 11 showed that more than 94 percent of Sacramento's red light violations occurred within 2 seconds after the onset of red light.

In this subsection, a logistic regression model is developed to illustrate the relationship between the dependent variable "elapsed time since red light onset" and a group of explanatory variables. The dependent variable, "RedEla2," has two categories of alternatives:

1. Drivers who ran through intersections when the elapsed time since red light onset is ≤ 2 seconds (SAS Code = 0), and
2. Drivers who ran through intersections when time elapsed since onset of red light is > 2 seconds (SAS Code =1).

Indecisive reaction when caught in "dilemma zone" could be one of the major reasons that drivers in the first group ran through the red light. When a driver is caught in the dilemma zone where he could not cross or stop comfortably, a delayed attempt to go through the intersection would trigger the red light photo enforcement camera. What could be some of the reasons that drivers still go through the intersection after the traffic light has turned red for more than 2 seconds? Some drivers might be intentionally violating the traffic law; however, there could be other drivers, especially older motorists, who might not react to the light change quick enough to stop their vehicles. Moreover, some drivers of all ages might be distracted or inattentive and thus unaware of the red light.

The explanatory variables used in this logistic regression analysis are same as those examined earlier in subsection 4.2.

Table 16 listed all available explanatory variables for this analysis. Two items should be noted on the explanatory variables:

- In the logistic analysis, "Time 4" (from 8 p.m. to 5 a.m.) was again selected as the reference variable during the data regression process. "Time 4" is the "off-peak" period for traveling with low traffic volume and red light violations in this time period are considerably lower compared to other hours of the day (see Figure 3).
- Variable codes "LocCode 1" to "LocCode 11" are the *dummy variables* representing the eleven Sacramento intersections used in the data analysis. "LocCode 10" (Howe Avenue and College Town Drive) is used as the reference variable for this logistic regression

analysis. "LocCode 10" is selected because this intersection has the highest clearance interval (5 seconds of yellow time + 2 seconds of all-red time, see Table 6) compared to the other ten intersections used in the analysis. As mentioned in a report by Eccles and McGee (2001), longer clearance interval will cause drivers to enter intersection later and generate disrespect for the traffic signal.

Using "RedEla2" as the dependent variable and all explanatory variables listed in Table 16, several runs of logistic regression were conducted so a fitting logit model that describes the relationship between "time elapsed since red light onset" and significant explanatory variables can be identified. Once again, necessary statistical tests were carried out to ensure assumed statistical properties are valid and the final model is robust.

After testing various combinations of explanatory variables, including some interaction terms such as "Age*Gender," a logit model was decided with "Age," "Time," and "LocCode" again as the explanatory variables. Main effects "Gender" and "VehYr" did not have any significant effect on "RedEla2" with p-values greater than 0.05 and were not included in the final model. None of the interaction terms tested showed significant influence on "RedEla2;" consequently, were not part of the final model.

Table 18 presents the estimated coefficients, standard errors, p-values, and odds ratios for all the explanatory variables in the final logit model where "RedEla2" is the dependent variable. Of the 46,997 red light violation records from Sacramento, a total of 39,045 records were used in this logistic regression analysis (7,952 records were not used because of missing values for either the dependent or the explanatory variables). The *generalized R^2* and *max-rescaled R^2* values for this model, both measuring the predictive power of the model, are 0.0456 and 0.1277 respectively.

Unlike the result for the "Speed2" model, the estimated coefficient for variable "Age" in the "RedEla2" model is a positive value. A positive coefficient indicates that the probability of running the red light increases as the age of the violator increases when the time elapsed since red light onset is greater than 2 seconds. The odds ratio for "Age" suggests that the predicted odds for an older driver running the red light when the time elapsed since red light onset is more than 2 seconds is about 1.2 times the odds of a middle-age driver.

At first glance, results for the "Age" variable may not be what we would normally expect since younger motorists tend to drive more aggressively and one would expect that a greater number younger drivers would run the red light when the time elapsed since red light onset is greater than 2 seconds compared to older drivers. However, researchers have found that older drivers require longer time to react to changes and their abilities to divide attention are not as good as younger drivers. Consequently, it is reasonable to see that the probability of an older driver running the red light when the time elapsed since red light onset is greater than 2 seconds would be higher than a younger driver, due to slower reaction time or unawareness of signal change when looking at the road ahead.

Table 18. Estimation Results from Binary Logit Model with "RedEla2" as Dependent Variable

Variable Code	Estimated Coefficient	Standard Error	p-value	Odds Ratio		1/Odds Ratio
Age	0.2139	0.0374	< 0.0001		1.238	0.807
Time 1	-1.3499	0.0756	< 0.0001	*versus Time 4*	0.259	3.857
Time 2	-1.6397	0.0567	< 0.0001	*versus Time 4*	0.194	5.153
Time 3	-1.8910	0.0688	< 0.0001	*versus Time 4*	0.151	6.626
LocCode 1	-0.8471	0.0688	< 0.0001	*versus LocCode 10*	0.429	2.333
LocCode 2	-1.0491	0.0804	< 0.0001	*versus LocCode 10*	0.350	2.855
LocCode 3	-1.5960	0.1742	< 0.0001	*versus LocCode 10*	0.203	4.933
LocCode 4	-1.2175	0.1218	< 0.0001	*versus LocCode 10*	0.296	3.379
LocCode 5	-1.9659	0.1597	< 0.0001	*versus LocCode 10*	0.140	7.141
LocCode 6	-1.2600	0.2337	< 0.0001	*versus LocCode 10*	0.284	3.526
LocCode 7	-2.2337	0.1268	< 0.0001	*versus LocCode 10*	0.107	9.334
LocCode 8	-0.8705	0.0763	< 0.0001	*versus LocCode 10*	0.419	2.388
LocCode 9	-0.4435	0.0704	< 0.0001	*versus LocCode 10*	0.642	1.558
LocCode 11	-2.1148	0.5843	0.0003	*versus LocCode 10*	0.121	8.288

The negative estimated coefficients for "Time 1," "Time 2," and "Time 3" suggest that drivers who run the red light in these three time periods have a lower probability of entering the intersection when the time elapsed since red light onset is more than 2 seconds compared to motorists who run the red light in "Time 4". The inverse odds ratios shown in Table 18 indicate that violators in "Time 4" are 3.9, 5.2, and 6.6 times more likely to run the red light later than 2 seconds since red light onset compared to violators in "Time 1," "Time 2," and "Time 3," respectively. These results are consistent with the results for the "Speed2" model. "Time 4," from 8 p.m. to 5 a.m., is the off-peak travel time, usually with relatively low traffic volumes. Total number of red light violations occurring during this time period is low, especially from midnight to 5 a.m. However, the situational factors in "Time 4," off-peak hours and light traffic, seem to create an atmosphere where more drivers are likely to run red light when the time elapsed since red light onset is more than 2 seconds and with vehicle speed higher than the posted speed limit.

Compared to the reference intersection, "LocCode 10," motorists who run the red light at "LocCode 1" to "LocCode 9" and "LocCode 11" all have a lower probability of entering these intersections when the time elapsed since red light onset is more than 2 seconds (negative estimated coefficients). Of the eleven Sacramento intersections studied in this report, "LocCode 10" has the highest yellow time at 5 seconds and all-red time at 2 seconds (see Table 6). In the field of traffic engineering, it is well recognized that when the yellow interval is too long, drivers will become accustomed to challenging the yellow time, enter the intersection later, and subsequently run the red light. The situation becomes even worse when there is an additional all-red phrase. Logistic regression results presented in this subsection support the belief of many traffic engineering professionals – "LocCode 10" has the longest clearance interval and most violators who ran the red light after 2 seconds of its onset.

The logit model for "RedEla2" presented in this subsection linked the effect of "Age," "Time," and "LocCode" to the time when violators enter the intersections after the onset of the red light. Results presented in this subsection gave us another level of understanding about red light running behavior. Implications of these results will be discussed in subsection 5.2.

5. CONCLUDING REMARKS

5.1. List of Major Findings

Notable findings from this study of Sacramento's red light violation data are highlighted below:

1. Most red light violations occurred during the normal work hours (i.e., 7 a.m. to 7 p.m.). Highest counts of red light violations occurred during the period from 2:00 p.m. to 2:59 p.m.
2. The average red light violation speed was 31.6 mph. More than half (51%) of the drivers ran the red light at speed of 30 mph or less. About 14 percent of the violators ran the red light at speeds higher than 45 mph.
3. More than 94 percent of the violations occurred within 2 seconds of red light onset.
4. As the age of the red light violator increases, the probability of running the red light while speeding decreases. The predicted odds of a younger driver running a red light while speeding is about 1.5 times the odds of a middle-age driver. The same odds ratio also applies when the age category goes from "middle-age" to "older."
5. Violators in time periods of "6 a.m. to 9 a.m.," "10 a.m. to 3 p.m.," and "4 p.m. to 7 p.m." have a lower probability of running the red light while speeding than those who commit red light violations in the time period of "8 p.m. to 5 a.m." Red light violators in this time period are 1.2, 2.4, and 2.0 times more likely to be speeding than those who run the red light in time periods "6 a.m. to 9 a.m.," "10 a.m. to 3 p.m.," and "4 p.m. to 7 p.m.," respectively. A combination of off-peak hours and low traffic volume may encourage violators to run the red light at speeds greater than the posted speed limit in the time period of "8 p.m. to 5 a.m."
6. Compared to the reference intersection for the "Speed2" logit model, "LocCode 4," violators at "LocCode 1," "LocCode 3," and "LocCode 9" have a lower probability of running the red light while speeding due to their heavy traffic volumes. Violators at remaining locations are more likely to run the red light while speeding than "LocCode 4." (See Table 16 for complete information regarding "LocCode 1" to "LocCode 11"). The probability of running the red light after 2 seconds of its onset increases as the age of the violator increases, according to the "RedEla2" logit model. The odds ratio for variable "Age" suggests that the predicted odds for an older driver running the red light after 2 seconds of its onset is about 1.2 times the odds of a middle-age driver. Older drivers require longer time to react to changes and their abilities to divide attention are not as good as younger drivers. Violators in time periods of "6 a.m. to 9 a.m.," "10 a.m. to 3 p.m.," and "4 p.m. to 7 p.m." have a lower probability of running the red light after 2 seconds of its onset than violators in the time period of "8 p.m. to 5 a.m." Violators in the time period of "8 p.m. to 5 a.m." are 3.9, 5.2, and 6.6 times more likely to run the red light after 2 seconds of its onset than violators in time periods of "6 a.m. to 9 a.m.," "10 a.m. to 3 p.m.," and "4 p.m. to 7 p.m.," respectively.
7. Most violations after 2 seconds of red light onset occurred at the intersection with the highest yellow plus all-red time.
8. Red light violation rates (calculated from issued violation citations) for the 11 Sacramento intersections ranged from 0.064 violation per 1,000 crossing vehicles to 0.294 violation per 1,000 crossing vehicles. In comparison to red light violation rates reported by other studies, these estimates of red light violation rates are quite low. A probable reason is because the red light violation records used to calculate these rates are actual citations issued and only approximately 35 percent of the photos taken by Sacramento's photo enforcement cameras lead to red light violation citations.

5.2. Implications of Findings

Implications of findings from this analysis are presented here from the perspective of designing and planning a CSVWS FOT:

1. The experimental design for the CSVWS FOT must examine the influence of driver age. Drivers in different age groups exhibit diverse behavior when approaching signalized intersections based on the findings of this study. How will older drivers respond to the warnings issued by CSVWS versus younger drivers? Will CSVWS warnings create different levels of distraction to drivers at different age groups?
2. Logistic regression analyses conducted in this study did not find significant relationship between red light violators' gender and their driving behavior approaching signalized intersections. Consequently, "gender" is a variable that could be excluded from the experimental design of the CSVWS FOT, especially when there are constraints on budget and time.
3. The logit models developed in this study suggest that the time period of "8 p.m. to 5 a.m." seems to have a combination of environmental and situational factors that would encourage red light violators to run through intersections at speeds higher than the posted speed limit or after the red light has elapsed for more than 2 seconds. When recruiting study participants for the CSVWS FOT, enough subjects must be included who would often travel in the off-peak period (e.g., 8 p.m. to 5 a.m.) so ample driver data from that time period can be gathered. Accordingly, comparison of driver reaction towards the CSVWS warnings at different time periods can be made.
4. Findings from the logit models presented in this report also suggest that several variations of the CSVWS warning algorithm and warning messages might be necessary for different time periods throughout the day. At certain time period(s) when drivers are susceptible of speeding through intersections or entering the intersections late when light changes, CSVWS warnings need to be issued earlier and warning messages need to be "decisive" to effectively encourage more drivers to stop for the red light.
5. Logistic regression results indicated that traffic volumes and duration of clearance intervals (yellow time and all-red phase) at intersections seem to have certain influence on red light violators' decisions to go through intersections at speeds greater than the posted speed limit or enter intersections when the elapsed time since red light onset is more than 2 seconds. In a CSVWS FOT, signalized intersections with various characteristics must be included in the experimental design so the effect of factors such as traffic volume and clearance interval on CSVWS can be closely studied.

6. REFERENCES

American City & County (2001). State Farm Names the 10 Most Dangerous Intersections. Available from www.americancityandcounty.com/news/government_state_farm_names/index.html.

BMI (December 2001). *Infrastructure-Based Intersection Collision Avoidance Concept Study, Technical Memorandum 2* (Contract No. DTFH61-96-C-00077). Silver Spring, Maryland: BMI.

Campbell, B. N., Smith, J. D., and Najm, W. G. (September 2004). *Analysis of Fatal Crashes Due to Signal and Stop Sign Violations* (DOT-VNTSC-NHTSA-02-09, DOT HS 809 779). Washington, DC: U.S. Department of Transportation.

Chang, M.-S., Messer, C. J., and Santiago, A. J. (1985). Timing Traffic Signal Change Intervals Based on Driver Behavior. *Transportation Research Record 1027*, 20-30. Transportation Research Board, Washington, DC.

Cerrelli, E. C. (January 1998). *Crash Data and Rates for Age-Sex Groups of Drivers, 1996* (Research Note, National Highway Traffic Safety Administration, U.S. Department of Transportation). Available from www.nhtsa.dot.gov/people/ncsa.

Brewer, M. A., Bonneson, J., Zimmerman, K. (2002). Engineering Countermeasures to Red-Light-Running. *Proceeding of the ITE 2002 Spring Conference and Exhibit (CD-ROM)*. Washington, DC: Institute of Transportation Engineers.

Eccles, K. A., and McGee H. W. (July 2001). *A History of the Yellow and All-Red Intervals for Traffic Signals*. Washington, DC: Institute of Transportation Engineers.

Fakhry, S. M. and Salaita, K. (2002). Aggressive Driving: A Preliminary Analysis of a Serious Threat to Motorists in a Large Metropolitan Area. *Journal of Trauma, 52(2)*, 217-224.

Federal Highway Administration (2002). *Highway Statistics 2001*. Washington DC: U.S. Department of Transportation.

Insurance Institute for Highway Safety (2000). Red Light Running Factors into More than 800 Deaths Annually; More than Half of Those Who Die are Hit by Red Light Violators. Available from www.iihs.org/news/2000/iihs_news_071300.pdf.

Kamyab, A., McDonald, T., and Stribiak, J. J. (2002). The Scope and Impact of Red Light Running in Iowa. *Preprint CD-ROM of the 81st Annual Meeting of the Transportation Research Board*, Washington, DC.

Kamyab, A., McDonald, T., Stribiak, J. J., and Storm, B. (December 2000). *Red Light Running in Iowa: The Scope, Impact, and Possible Implications*. Ames IA: Center for Transportation Research and Education, Iowa State University.

Kraus, E. and Quiroga, C. (2004). Red Light Running Trends in Texas. *Preprint CD-ROM of the 83rd Annual Meeting of the Transportation Research Board*, Washington, DC.

Lum, K. M. and Wong, Y. D. (2003). Impacts of Red Light Camera on Violation Characteristics. *Journal of Transportation Engineering, November/December,* 648-656.

McGee, H. W. (2002). Safety Impact of Red Light Camera Enforcement Program. *Proceeding of the ITE 2002 Spring Conference and Exhibit (CD-ROM).* Washington, DC: Institute of Transportation Engineers.

Milazzo II, J. S., Hummer, J. E., and Prothe, L. M. (June 2001). *A Recommended Policy for Automated Electronic Traffic Enforcement of Red Light Running Violations in North Carolina.* Raleigh, North Carolina: Institute for Transportation Research and Education, North Carolina State University.

Milazzo II, J. S., Hummer, J. E., Rouphail, N. M., Prothe, L. M., and McCurry, J. B. (2002). The Effect of Dilemma Zones on Red Light Running Enforcement Tolerances. *Preprint CD-ROM of the 81st Annual Meeting of the Transportation Research Board,* Washington, DC.

Porter, B. E. and Berry, T. D. (2001). A Nationwide Survey of Self-Reported Red Light Running: Measuring Prevalence, Predictors, and Perceived Consequences. *Accident Analysis and Prevention, 33,* 735-741.

Porter, B. E. and England, K. J. (2000). Predicting Red-Light Running Behavior: A Traffic Safety Study in Three Urban Settings. *Journal of Safety Research, 31(1),* 1-8.

Retting, R. A., Chapline, J. F., and Williams, A. F. (September 2000). *Changes in Crash Risk Following Re-Timing of Traffic Signal Change Intervals.* Arlington, Virginia: Insurance Institute for Highway Safety.

Retting, R. A. and Green, M. A. (1997). Influence of Traffic Signal Timing on Red-Light Running and Potential Vehicle Conflicts at Urban Intersections. *Transportation Research Record 1595,* 1-7. Transportation Research Board, Washington, DC.

Retting, R. A., and Kyrychenko, S. Y. (April 2001). *Crash Reductions Associated with Red Light Camera Enforcement in Oxnard, California.* Arlington, Virginia: Insurance Institute for Highway Safety.

Retting, R. A., and Kyrychenko, S. Y. (2002). Reductions in Injury Crashes Associated with Red Light Camera Enforcement in Oxnard, California. *Preprint CD-ROM of the 81st Annual Meeting of the Transportation Research Board,* Washington, DC.

Retting, R. A. and Williams, A. F. (1996). Characteristics of Red Light Violators: Results of a Field Investigation. *Journal of Safety Research, 27(1),* 9-15.

Retting, R. A., Ulmer, R. G., and Williams, A. F. (1999). Prevalence and Characteristics of Red Light Running Crashes in the United States. *Accident Analysis and Prevention, 31,* 687-694.

Retting, R. A., Williams, A. F., Farmer, C. M., and Feldman, A. F. (1999a). Evaluation of Red Light Camera Enforcement in Oxnard, California. *Accident Analysis and Prevention, 31,* 169-174.

Retting, R. A., Williams, A. F., Farmer, C. M., and Feldman, A. F. (1999b). Evaluation of Red Light Camera Enforcement in Fairfax, Virginia. *ITE Journal, Vol. 69, No. 8,* 30-34.

Retting, R. A., Williams, A. F., and Greene, M. A. (1998). Red-Light Running and Sensible Countermeasures: Summary of Research Findings. *Transportation Research Record 1640,* 23-26. Transportation Research Board, Washington, DC.

Retting, R. A., Williams, A. F., Preusser, D. F., and Weinstein, H. B. (1995). Classifying Urban Crashes for Countermeasure Development. *Accident Analysis and Prevention, 27(3),* 283-294.

Ruby, D. E. and Hobeika, A. G. (2003). Assessment of Red Light Running Cameras in Fairfax County, Virginia. *Transportation Quarterly, Vol. 57, No. 3,* 33-48.

Schattler, K. L., Hill, C., and Datta, T. K. (2002). Clearance Interval Design and Red Light Violations. *Proceeding of the ITE 2002 Spring Conference and Exhibit (CD-ROM).* Washington, DC: Institute of Transportation Engineers.

Zador, P., Stein, H., Shapiro, S., and Tarnoff, P. (1985). Effect of Signal Timing on Traffic Flow and Crashes at Signalized Intersection. *Transportation Research Record 1010,* 1-8. Transportation Research Board, Washington, D..

Appendix A. Photos of City of Sacramento's 11 RLPEC-Equipped Intersections

Figure 12. Picture of Fair Oaks Boulevard and Howe Avenue Intersection

Figure 13. Picture of El Camino Avenue and Evergreen Street Intersection

Figure 14. Picture of Arden Way and Exposition Boulevard Intersection

Figure 15. Picture of Mack Road and La Mancha Way/Valley Hi Drive Intersection

Figure 16. Picture of Mack Road and Center Parkway Intersection

Figure 17. Picture of 30th Street and Capitol Avenue Intersection

Figure 18. Picture of J Street and Alhambra Boulevard Intersection

Figure 19. Picture of Broadway and 21st Street Intersection

Figure 20. Picture of W Street and 16th Street (Highway 50 Exit at 16th Street) Intersection

Figure 21. Picture of Howe Avenue and College Town Drive Intersection

Figure 22. Picture of Power Inn Road and Folsom Boulevard Intersection

Appendix B. Photos of Red Light Photo Enforcment Camera and Warning Sign

Figure 23. Picture of Red Light Photo Enforcment Camera in the City of Sacramento

Figure 24. Picture of Red Light Photo Enforcment Camera Warning Sign in the City of Sacramento

Appendix C. GIS Map of a Selected RLPEC-Equipped Intersection in Sacramento

Figure 25. GIS Mapping of Fair Oaks Boulevard and Howe Avenue Intersection

Appendix D. Distributions of Red Light Violations by Speed at Individual 11 RLPEC-Equipped Intersections

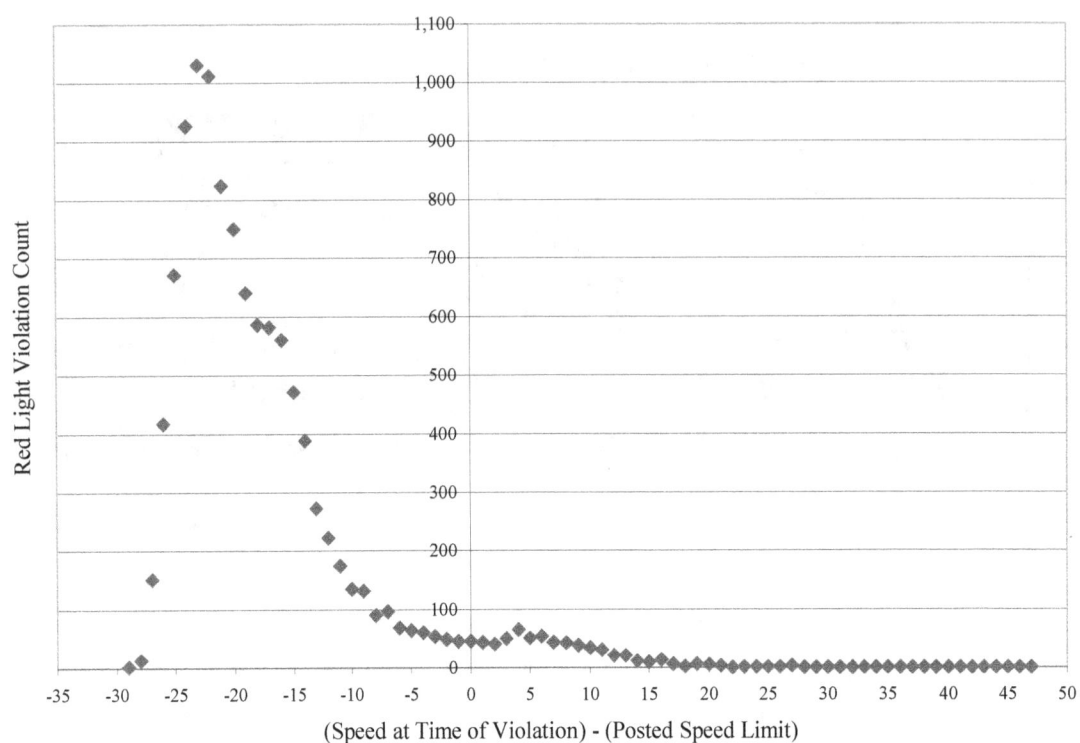

Figure 26. Frequency Distribution of Red Light Violations by Speed at Fair Oaks Boulevard and Howe Avenue Intersection

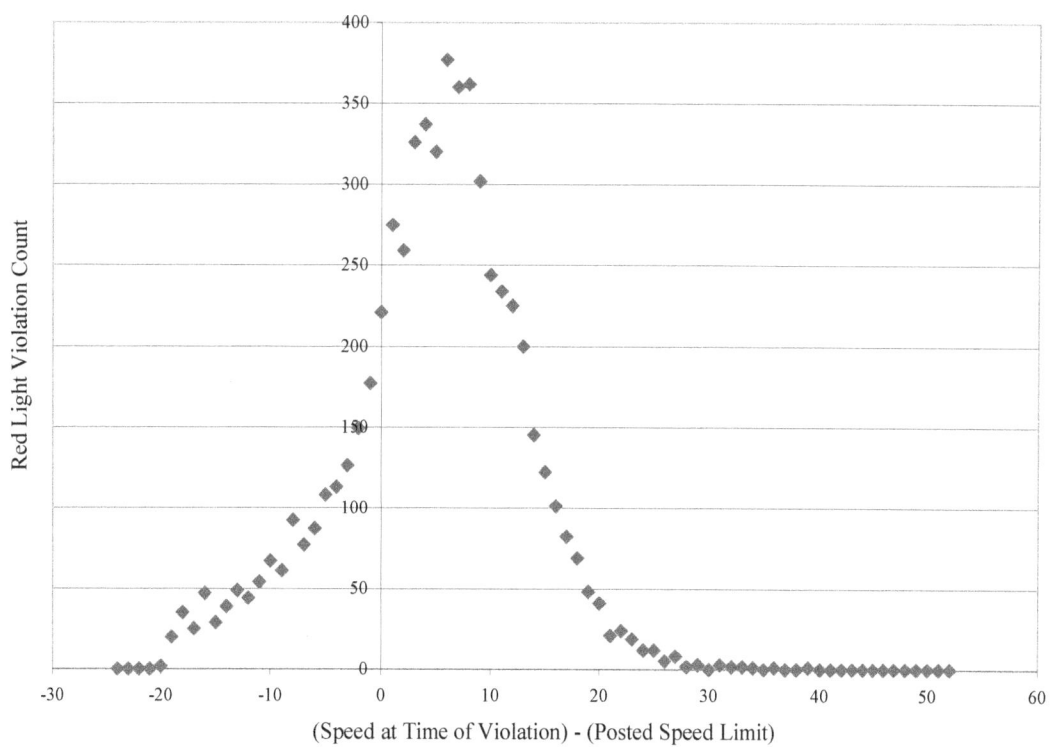

Figure 27. Frequency Distribution of Red Light Violations by Speed at El Camino Avenue and Evergreen Street Intersection

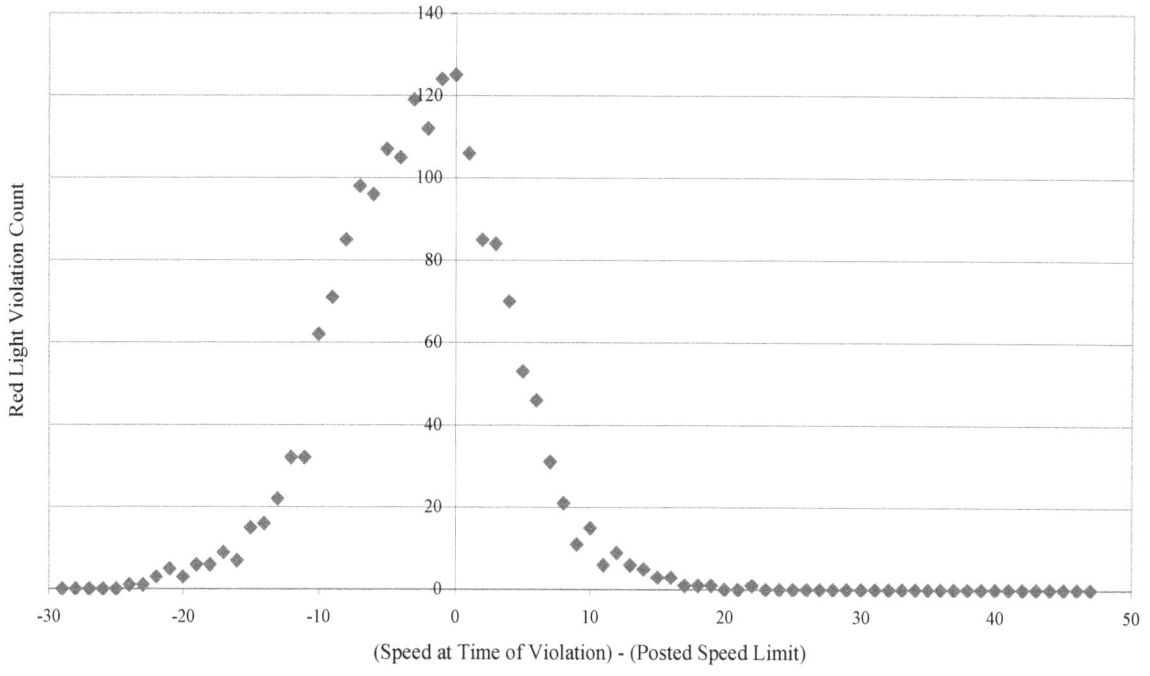

Figure 28. Frequency Distribution of Red Light Violations by Speed at Arden Way and Exposition Boulevard Intersection

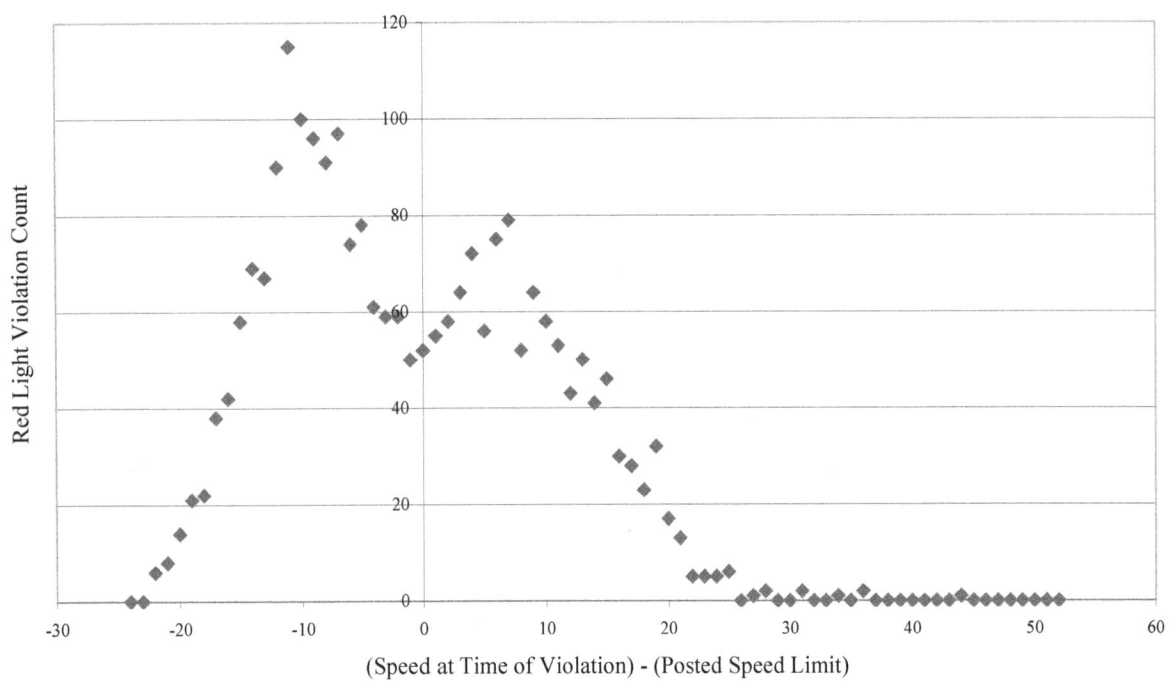

Figure 29. Frequency Distribution of Red Light Violations by Speed at Mack Road and La Mancha Way/Valley Hi Drive Intersection

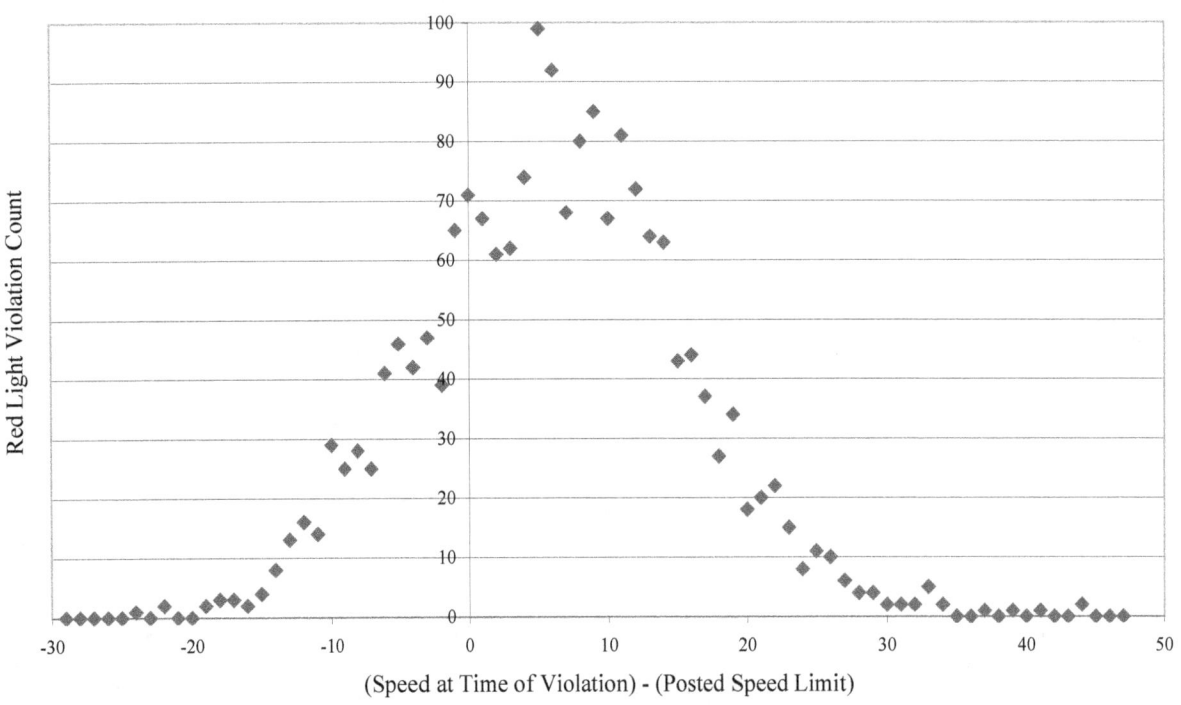

Figure 30. Frequency Distribution of Red Light Violations by Speed at Mack Road and Center Parkway Intersection

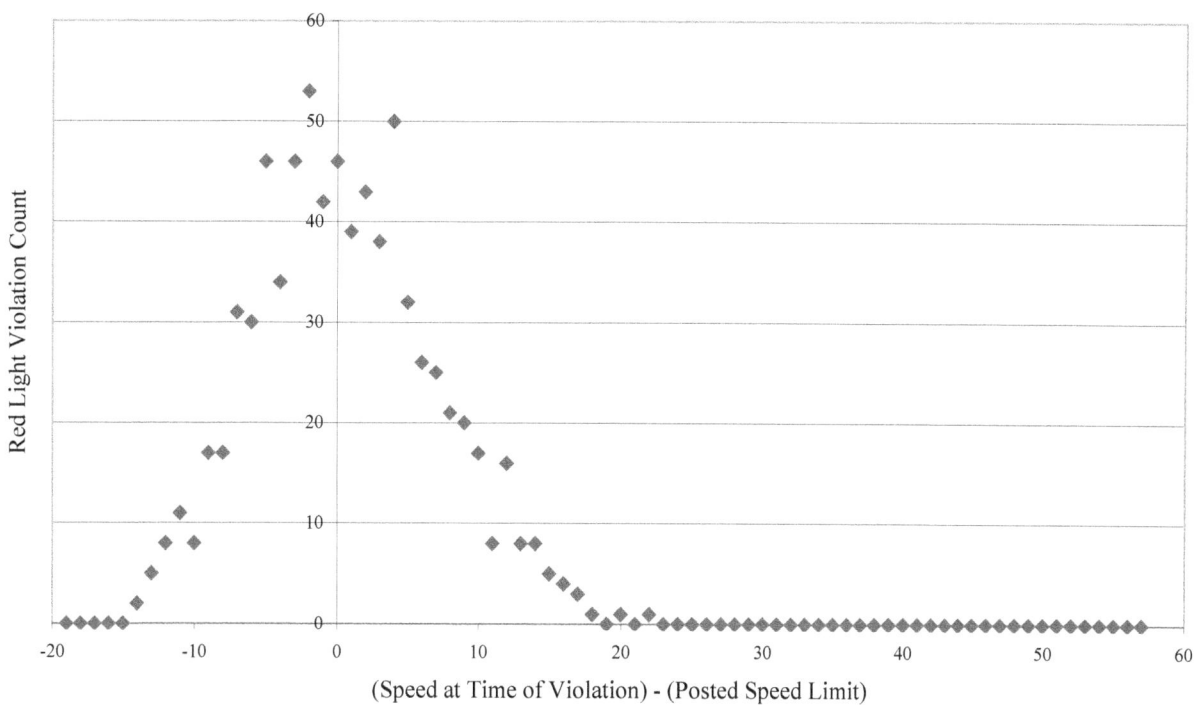

Figure 31. Frequency Distribution of Red Light Violations by Speed at 30th Street and Capitol Avenue Intersection

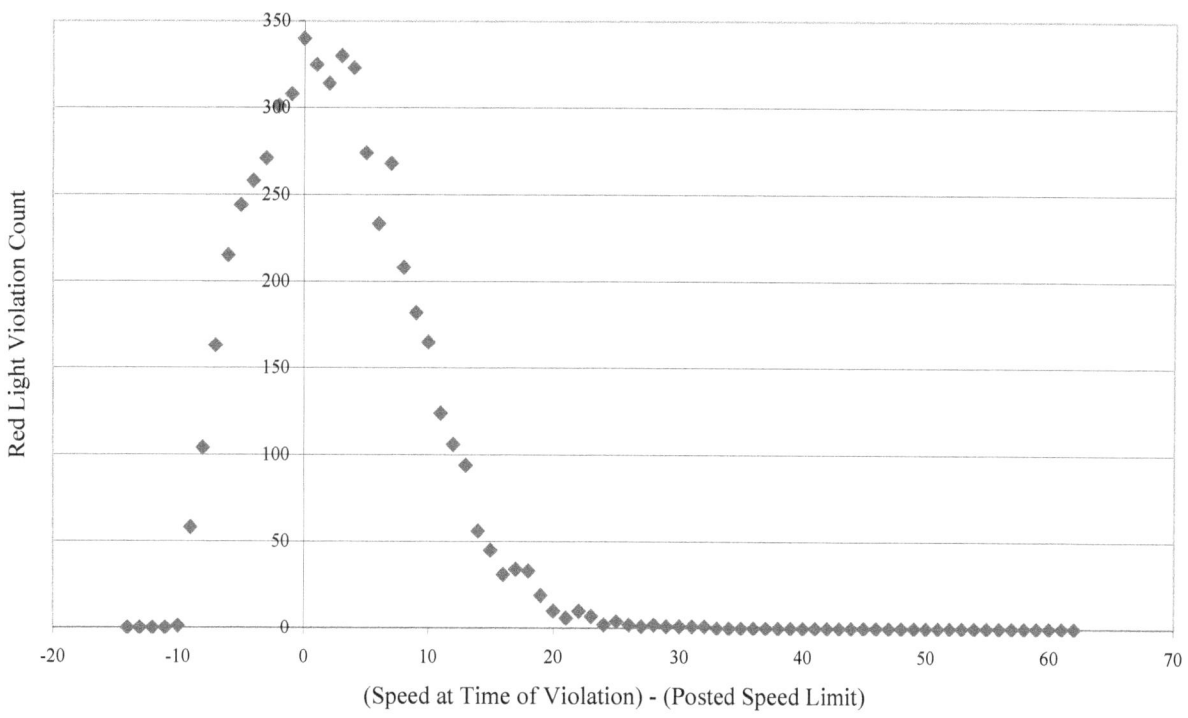

Figure 32. Frequency Distribution of Red Light Violations by Speed at J Street and Alhambra Boulevard Intersection

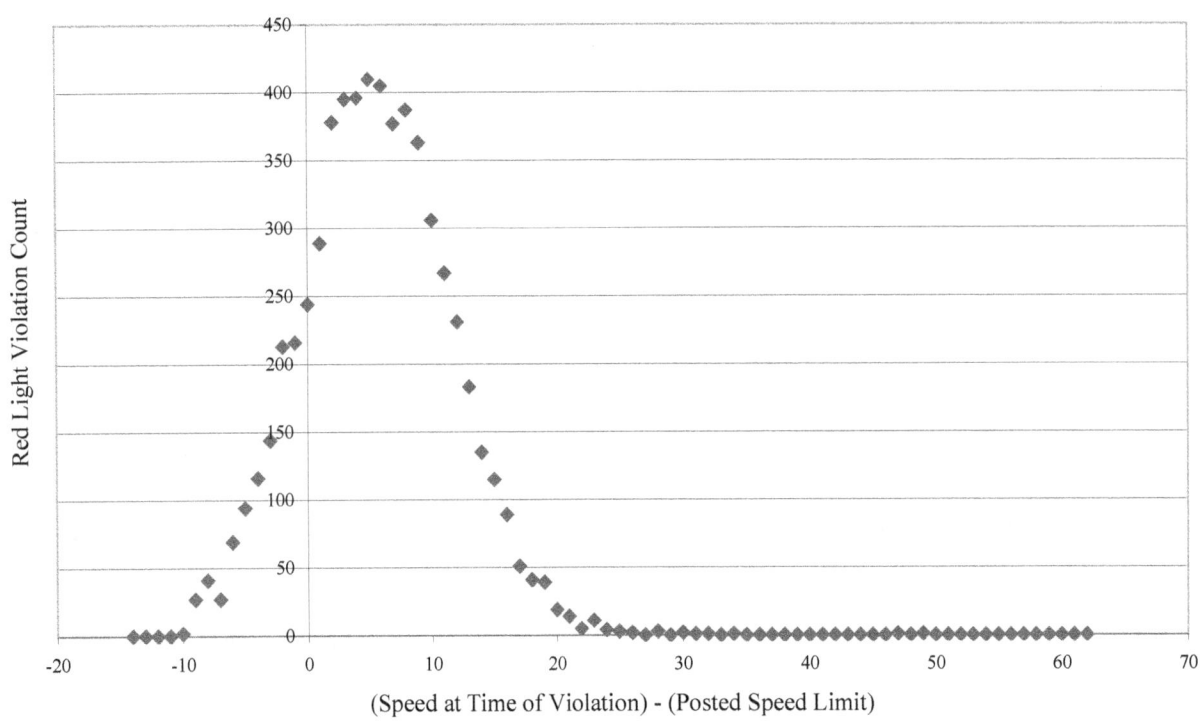

Figure 33. Frequency Distribution of Red Light Violations by Speed at Broadway and 21st Street Intersection

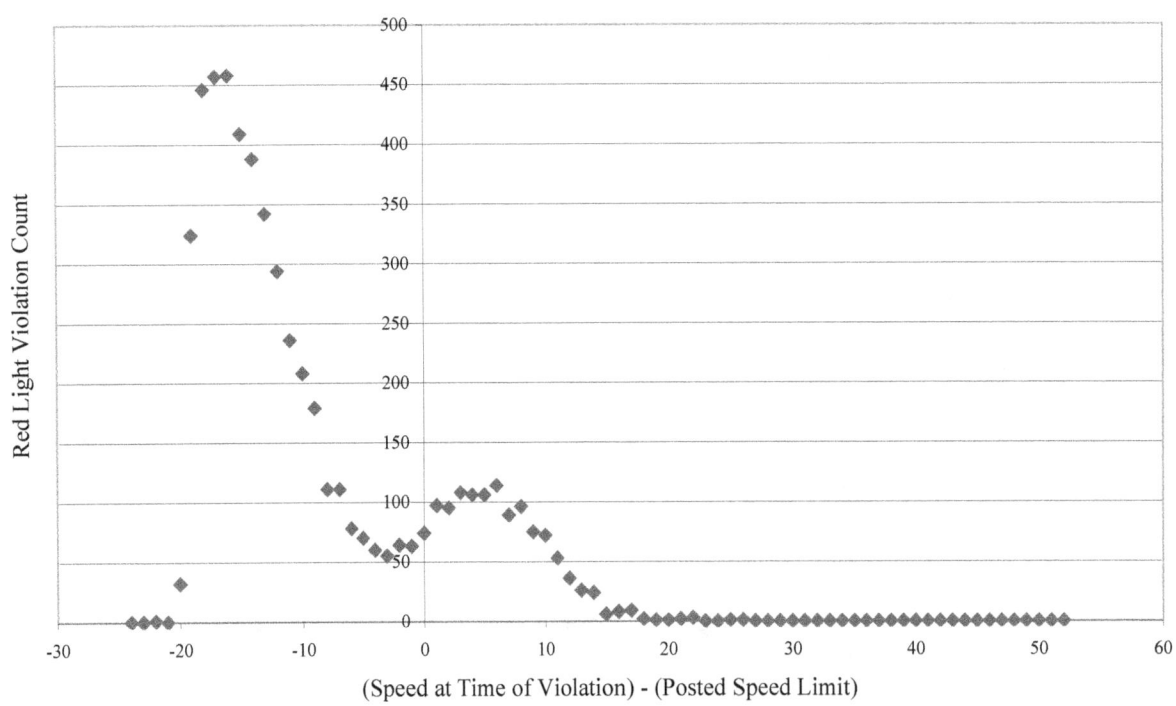

Figure 34. Frequency Distribution of Red Light Violations by Speed at W Street and 16th Street (Highway 50 Exit at 16th Street) Intersection

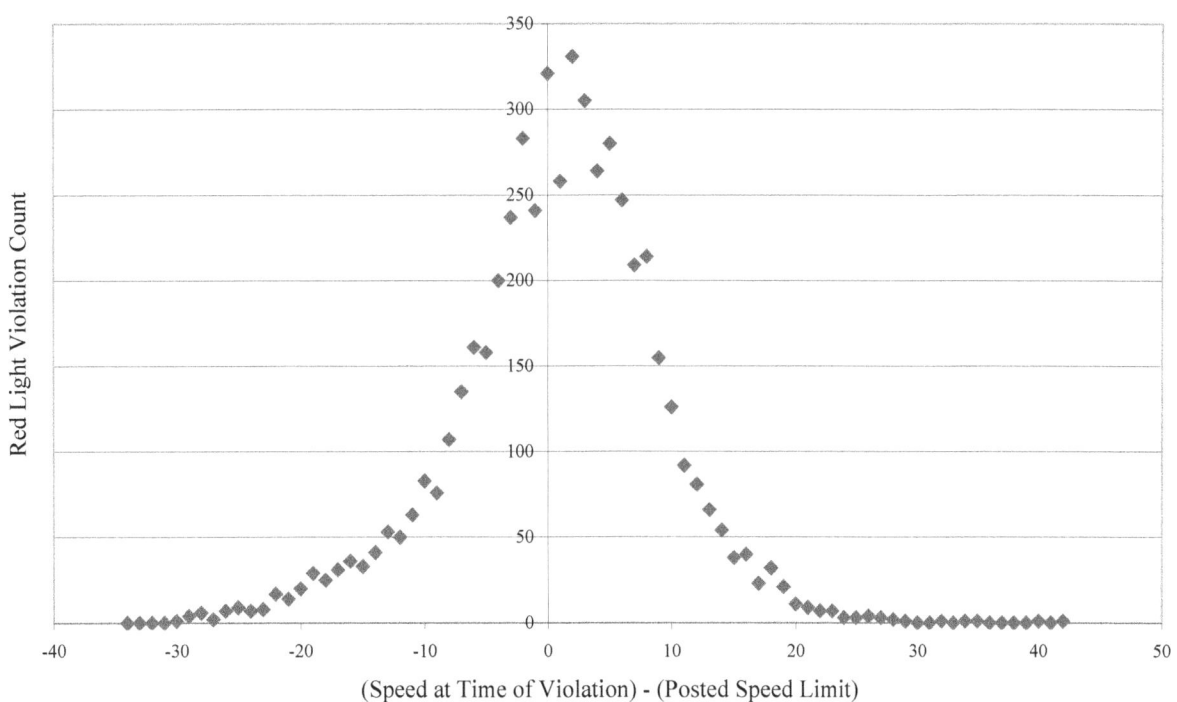

Figure 35. Frequency Distribution of Red Light Violations by Speed at Howe Avenue and College Town Drive Intersection

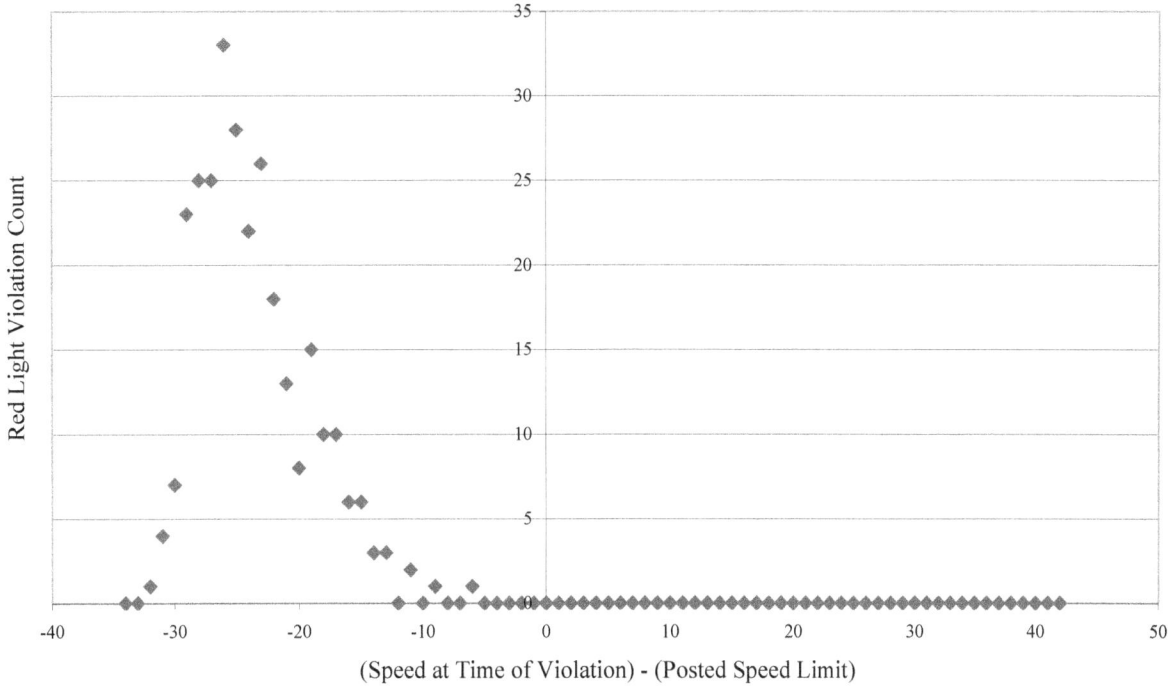

Figure 36. Frequency Distribution of Red Light Violations by Speed at Power Inn Road and Folsom Boulevard Intersection

DOT-VNTSC-NHTSA-05-01
DOT HS 810 580
March 2006

U.S. Department
of Transportation

**National Highway
Traffic Safety
Administration**

Research and
Innovative Technology
Administration
Volpe National
Transportation System Center
Cambridge, MA 02142-1093

www.ingramcontent.com/pod-product-compliance
Lightning Source LLC
Chambersburg PA
CBHW081848170526
45167CB00007B/2929